The Idea of Technological Innovation

To Joseph P. Lane, friend, colleague
and enlightened practitioner

Technological innovation is only one route to
growth . . . Scientific research is only one route to
technological innovation.
(Donald Schon, *Technology and Change: The New
Heraclitus*, 1967)

The Idea of Technological Innovation

A Brief Alternative History

Benoît Godin

Institut National de la Recherche Scientifique (INRS), Canada

Edward Elgar
PUBLISHING

Cheltenham, UK • Northampton, MA, USA

Published by
Edward Elgar Publishing Limited
The Lypiatts
15 Lansdown Road
Cheltenham
Glos GL50 2JA
UK

Edward Elgar Publishing, Inc.
William Pratt House
9 Dewey Court
Northampton
Massachusetts 01060
USA

A catalogue record for this book
is available from the British Library

Library of Congress Control Number: 2019956684

This book is available electronically in the **Elgar**online
Social and Political Science subject collection
DOI 10.4337/9781839104008

ISBN 978 1 83910 399 5 (cased)
ISBN 978 1 83910 401 5 (paperback)
ISBN 978 1 83910 400 8 (eBook)

Typeset by Servis Filmsetting Ltd, Stockport, Cheshire
Printed and bound in Great Britain by TJ International Ltd, Padstow

Contents

Introduction

Over the last few decades, technological innovation became the new religion of our society, the modern belief or faith. Innovation is a *panacea* for our socioeconomic problems:

> Most current social, economic and environmental challenges require creative solutions based on innovation and technological advance. (OECD, 2010: 30)

> Innovation is our best means of successfully tackling major societal challenges, such as climate change, energy and resources scarcity, health and ageing, which are becoming more urgent by the day. (European Commission, 2010: 2)

Such a belief has a long history. "There is little doubt", stated the OECD in one of the first titles on technological innovation ever produced in the western world, that:

> If governments succeed in helping to increase the pace of technical innovation, it will facilitate structural changes in the economy, and increase the supply of new and improved products necessary for Member Governments to achieve rapid economic growth and full employment and without inflation. (OECD, 1966: 8)

> Do we need to innovate? ... Yes because it is one way, perhaps one of the best ways, to react in our rapidly changing society. (OECD, 1969: 15)

1

One could look further back in time for similar discourses on the broader concept of innovation (rather than technological innovation) and its use from the late nineteenth century onward (Godin, 2015). Yet, innovation definitely acquired a dominant and positive value thanks to or because of technological innovation. Innovation came to be what it is to us now in the context of technological innovation.

Technological innovation is a term that emerged after the Second World War, with only a few uses before that time (See Hansen, 1932; Kuznets, 1929: 540; Schumpeter, 1939: 289; Stern, 1927, 1937; Usher, 1929: vii, 10; Veblen, 1915: 118, 128–9) (see Figure I.1). Joseph Schumpeter is often credited with introducing the concept of innovation into economics. But the concept was used regularly among statisticians, economists and economic historians to discuss technology during Schumpeter's time.

How did technological innovation come to be an object of imagination and imaginaries? The answer lies deep in the 1950s. On one side, people started producing thoughts on what innovation is, how it happens, and with what effects. Economic growth – "growthmanship" as some called it – supported by public policy, gave the concept of technological innovation a social existence. On the other side, policymakers started inventing policies and strategies to support innovation, thus legitimizing the emerging discourse.

The historiography of technological innovation

as a term or concept and as a discourse is generally told from a theoretical point of view. The story goes like this: economist Joseph Schumpeter was the precursor of the concept in the 1930s–40s. The literature exploded in the 1960s–70s, a time at which the British Science Policy Research Unit (SPRU) played a key role. In between, some researchers made theoretical contributions (for example, Robert Solow), particularly to the argument for public support of scientific research (for example, Kenneth Arrow). To some extent, such a story has germs of truth. These dates are moments in the development of the discourse on technological innovation. But this is far from the whole story.[1] As mentioned above, Schumpeter was just one among several authors who wrote on innovation at the time. Moreover, Schumpeter's influence occurred much later – although he was influential on the literature on "technological change" early on (Godin, 2019).

The above illuminates how academics construct their own history. This book develops an alternative historiography and, following historian Ann Johnson on applied research, I ask: what if we wrote the history of the technological innovation from another perspective (Johnson, 2008)? I suggest here that practitioners (engineers, managers, policymakers and their advisers and consultants) have been pioneering theorists of technological innovation, beginning in the 1950s. It is the practitioners' view that scholars articulated later on and theorized about.

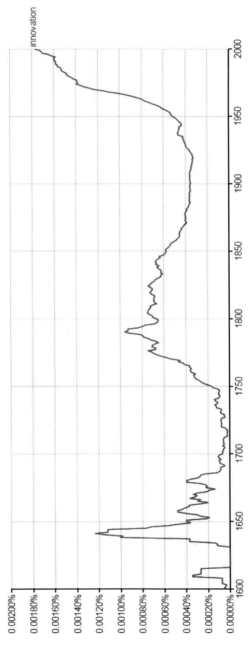

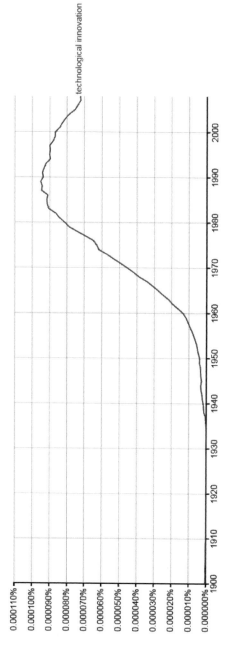

Figure I.1:   Frequency of the terms "innovation" and "technological innovation" over time (Google Ngram)

To appreciate this, one has to deconstruct the standard historiography of innovation. The actual historiography mixes the history of the idea, or concept of innovation with that of invention, the history of technology with that of science, the history of development with that of research. Hence the acronyms R&D (research and development), ST (science and technology), then STI (science, technology and innovation) that serve as umbrellas. However, as John Jewkes, Professor of Economics at Merton College, Oxford, put it long ago: "It has become fashionable to speak of 'science and technology' in the singular as if what was true of one was inevitably true of the other. But the generalizations that can be made about the two in common are few" (Jewkes, 1960: 1).[2] Similarly, a little later David Novick of RAND Corporation suggested: "We should stop talking about research and development as though they were an entity and examine research on its own and development as a separate and distinct activity" (Novick, 1965: 13; see also Novick, 1960).

Once these concepts are distinguished, we get a totally different story. This book revises the standard historiography of technological innovation. I document how, in a first step, one concept (research) gave rise to another (innovation) and how, in a second step, research got marginalized in the discourses on technological innovation. These two steps or stages correspond to two discourses, espoused by two different communities. One discourse postulates that innovation results from the application of science to industry. The issue at the heart of this discourse is research and development (R&D) and scientists and engineers, as assets to a country's competitiveness. Here, innovation is an article of faith (the ultimate outcome emerging out of basic research), and not really theorized about. This discourse is still popular today. For example, R&D remains one of the key measurements of innovation and a key variable in models of innovation. The second discourse regards innovation as a complex activity that, if it refers to research at all, only does so as one part of a whole process. The most important part or step of this process is not research, but the development of products and their commercialization. Although originating from the same conceptual framework (growthmanship), research and innovation are opposite values.

I organize the history of the concept of technological innovation around five phases, from 1950 to the present. The phases are not strictly discrete and linear, but overlap:

1. Innovation as science applied
2. Innovation as outcome
3. Innovation as process
4. Innovation as system
5. Innovation policies

This book is the result of more than 20 years of research. It summarizes and develops further ideas originally documented in previous books

(Godin, 2005, 2017, 2019). Over the course of my studies, I realized that the story of innovation as an idea is told from a strictly scholarly point of view, as though technological innovation as a concept and the study of innovation emerged and developed in academia. I thus started to seriously study the material from other sources that I had collected over the years. This led to the present book.[3]

To some, the documents on which the book rests will appear to include a large quantity of Anglo-Saxon sources. Right. The history of the concept of innovation over the centuries leads precisely to the predominant role of the United States in the twentieth century. The concept of innovation entered our everyday vocabulary in England during the Reformation. For the following three centuries, the concept had a predominantly nega-tive and pejorative connotation (Godin, 2015). In the nineteenth century, the concept traveled to France where, after the Revolution, it gradu-ally acquired a more positive value. Innovation is an instrument of material, social and political progress. Lately, in the twentieth century, people made of the concept a technological affair, with superlative overtones – a discourse that is more and more contested. The United States is a major contributor to this representation.

This book looks at the times when and the places where practitioners as thinkers and writers can be considered precursor theorists of technologi-cal innovation. The United States after the Second

World War is such a place. Britain entered the field at the same time. Practitioners from other countries followed, under the influence of international organizations such as the OECD. The book gives equal place to all these contributors, to the extent that they have been "innovative ideologists", to use intellectual historian Quentin Skinner's term, redescribing the world in a new moral light (Skinner, 2002a, 2002b).

Notes

1. The story does not distinguish between the literature on research (or science) and that on innovation, hence the tendency to include a diversity of authors not concerned directly with innovation (like Arrow). The distinction between the two concepts is discussed at length in the present book.
2. See also Burns and Stalker, 1961: 156–73; Jewkes et al., 1958: 17–25, 197–222; Rosenberg, 1976.
3. The history documented in this book should not be interpreted as a marginalization of the scholarly contribution to the study of technological innovation. I wish to offer an additional perspective for those who want to understand the sources of ideas in culture and society.

1. Prehistory

The discourse on technological innovation did not emerge from nothing. It emerged from, and as a reaction to, the previous discourses and concepts on which it is constituted. One such concept is technology (a good that embodies knowledge). Technological innovation is the use of technology in practice, not the technology itself – although it is often used in that sense. Second, technology gave rise to a plethora of concepts with the adjective "technological" in their titles. Before technological innovation came on the scene, several other concepts or terms paved the way for a broader understanding of technology, namely the consideration of its effects on society: technological unemployment, technological change and technological progress.

Technological unemployment

At the heart of the technology question at the beginning of the twentieth century was labor issues. The first term coined with technology in its name, or at the least debated extensively, was technological unemployment. The term emerged in the context of the depressions of the 1920s. Following the crash of the 1920s, "the term 'technological unemployment', first coined by economist Sumner Slichter in the second half of the 1920s (Slichter, 1928), became current", as economist Harry Jerome of the National Bureau of Economic Research (NBER) stated in *Mechanization in Industry* (1934). "Then, with the close of the decade, a prolonged depression set in . . . There came a sharp emphasis upon, if not an exaggeration of, the actual and potential adverse effects of the machine" (Jerome, 1934: 4–5). Technological unemployment is unemployment due to the introduction of "machines" or "inventions" in industry – two key terms of the nineteenth century. Until then, at least to classical economists, unemployment due to machines or inventions was reabsorbed automatically, so it was said (see Gourvitch, 1940). Then, in the early 1930s, the economists added "time" to the equation. The question was then whether technological unemployment is permanent or temporary.

The debate ended with a relative but dominant consensus. Technological unemployment is temporary. It is a matter of job displacements. A job lost somewhere is gained elsewhere. "For every man laid off a new job has been created somewhere", claimed Paul Douglas, a major theorist of the time (Douglas, 1930: 930). Not all inventions cause unemployment. Some create totally new industries and new types of employment. Technological unemployment depends on types of invention (labor-saving or capital-saving) and types of job affected (skilled, semi-skilled, unskilled).

By the 1950s, few doubted that inventions are beneficial to the economy:

> Not a single witness [as the congressional Subcommittee on Economic Stabilization recounted in a report on automation and technological change] raised a voice in opposition to automation and advancing technology. This was true of the representatives of organized labor as well as of those who spoke from the side of management. None of the evidence available before the subcommittee supports a charge that organized labor opposes or resists dynamic progress. (US Congress, 1955: 4–5)

> I did not once hear during the session the words "technological unemployment" [claimed Gerhard Colm, Chief Economist at the National Planning Association, to the participants at a National Science Foundation conference on research and development and its impact on the economy]. It is very remarkable that in the midst of a serious recession, with a large number of unemployed, no one suggested that there is any relationship between technological advance and unemployment. For the present period, I do not think this is an omission ... Research is adding to demand and thereby creating more jobs than it is displacing. (Colm, 1958: 152)

Technological change

The term technological change emerged out of the debate on technological unemployment. The term was coined in the 1920s (Alford, 1929; Hansen, 1921; Kuznets, 1929, 1930). It entered the vocabulary widely in the late 1930s. The National Research Project (NRP) on "Reemployment Opportunities and Recent Changes in Industrial Techniques" of the Works Progress Administration (WPA) is a key factor here. Between 1937 and 1940, the NRP conducted surveys of workers' experiences with machines in industry, with over 1,000 assistants in the field (Weintraub and Kaplan, 1938). Under the direction of David Weintraub, a young economist from the National Bureau of Economic Research (NBER), the NRP produced dozens of studies in collaboration with the NBER and several government departments and agencies (Department of Agriculture, Bureau of Labor, Bureau of Mines). To the NRP, technological change is change in industrial techniques of production, and it is measured like technological unemployment is measured: output per man-hour (Weintraub, 1932). Changes in productivity indicate changes in industrial techniques.

To the NRP, technological change is a source of positive effects on society (outputs, productivity, economic progress), not just negative ones (unemployment). But in order to realize the full benefits of technological change on the economy, there is a need for more investment in capital (industrial techniques). Here, the NRP was espousing economist Simon Kuznets's retardation thesis and Alvin Hansen's secular stagnation theory: investments in capital were more numerous and revolutionary in the past than they were at that time, hence the need for more investments (Hansen, 1939; Kuznets, 1929, 1930). This is a matter of business cycles. Joseph Schumpeter explained periods of recession and periods of prosperity based on the spread of such revolutionary innovations (Schumpeter, 1939).

From that time on, technological change became a catchword whose meaning includes (or oscillates between) change *in* industrial techniques (the technical) and the changes to society and the economy (the change) *due to* industrial techniques. The key use of the term is the latter. The study of change in industrial techniques per se is left to historians (of invention).

In the 1940s technological change acquired a more restricted meaning that became very influential among economists. Defined by way of what is called the production function (combining input – labor and capital – to produce output) (Brozen, 1942; Schumpeter, 1939), the study of technological change became an industry. Econometric studies were published by the dozen in the economic literature. Certainly, other representations of technological change existed. To sociologists, anthropologists and philosophers, technological change is broader than just the use of industrial techniques in production. It includes any invention to which people have to adjust: farm practices, transportation, communication, and so on. But these disciplines did not develop a discourse on technological change per se. For example, the US Social Science Research Council, supported by sociologist of invention William Ogburn, tried to get into the field in the late 1940s, but it failed in its efforts (Godin, 2019). By that time, technological change had an economic (and mathematical) connotation that made it hard for sociologists to follow.

By the 1950s–60s, technological change got on to the public agenda and became "a modern sounding term", as one of the chapters of the US Commission on National Goals put it (Watson, 1960: 193). Organizations like the OECD, UNESCO and the International Labour Office, as well as policymakers and advisers, espoused the concept (together with that of automation) and developed thoughts for policies on technological change.[1] By that time, technological unemployment was no longer a fundamental issue. "Technological advance is a key element in economic progress" ("increases in our standard of living"), as the US Council of Economic Advisers put it, "Our objective should be to foster and encourage it" (US Council of Economic Advisers, 1964: 103, 85).

Technological progress

Practitioners and scholars also began to suggest that a key factor or condition of technological change – namely, science – has not been considered enough. Science is the source of technological progress. The concept of technological progress is not easy to define. The meaning oscillates between progress *in* technology (industrial techniques) and economic progress *due to* technology (growth in output) (Fabricant, 1965: 5–6; Kennedy and Thirlwall, 1972: 12). The two are often considered to be synonymous. "Economic progress which results from a change in knowledge is known as technological progress" claims Vernon Ruttan, economist at the University of Minnesota and a prolific scholar on technological change in the 1950s and after (Ruttan, 1954: 1).

The first to espouse a discourse on science and progress in the twentieth century were US industrialists such as Frank Jewett and John Carty in the 1920s, supported by the US National Research Council (Godin, 2009; Godin and Schauz, 2016). The Department of Scientific and Industrial Research (DSIR) in Britain held a similar discourse (British Department of Scientific and Industrial Research, 1927). Subsequently, a whole literature developed on the management of industrial research. The discourse culminated in Vannevar Bush's *Science: The Endless Frontier*, which proposed the creation of a National Research Foundation to support basic research publicly (Bush, 1945):

> Advances in science ... when put to practical use ... insure our health, prosperity, and security as a nation in the modern world. (p. 11)

> Basic research ... creates the fund from which the practical applications of knowledge must be drawn. New products and new processes do not appear full-grown. They are founded on new principles and new conceptions, which in turn are painstakingly developed by research in the purest realm of science ... Basic research is the pacemaker of technological progress. (p. 19)

To make sense of the discourse, industrialists developed models, as we call them today, picturing technological progress in terms of a sequence called "diagram" (Mees, 1920), or a "research cycle"

(Holland, 1928), or "flow chart" (Bichowsky, 1942; Furnas, 1948), or simply "stages" (Stevens, 1941) (see Figure 1.1). These sequences culminated in that of economic historian Rupert Maclaurin. In the 1940s, Maclaurin from MIT, developed the first-ever scholarly research program on technological change – over the years, Maclaurin shifted to the term "technological innovation". Maclaurin gave central place to research as a factor of technological change, explaining technological change as a "sequence" from research to use: pure science, invention, innovation, finance, acceptance (Maclaurin, 1949, 1953). By the 1960s, some called this sequence the rational or "common-sense view" (Schon, 1967: 54, 140). Subsequently, the sequence came to be called the "linear model of innovation" (Langrish et al., 1972; Price and Bass, 1969).

In the late 1950s and early 1960s, reviews (see, for example, Nelson, 1959a), conferences (see US National Bureau of Economic Research, 1962) and textbooks on the economics of science (see Bright, 1964) began to appear. In the end, most scholars were concerned with research (as source of or input to economic progress), not innovation (as output of research activities, which is taken for granted). "This [review] paper", claimed Richard Nelson of RAND Corporation, "is not concerned with innovation. It is concerned only with how inventions occur" (Nelson, 1959a; see also National Bureau of Economic Research, 1962: 4). Nelson discusses but criticizes and minimizes innovation issues (the "overemphasis on the market or 'demand' side" of invention; social needs explaining invention) espoused by sociologists like William Ogburn and Colum Gilfillan, and economists like Jacob Schmookler, at the cost of de-emphasizing research. What motivated Nelson to make such a move? The belief in basic research as a prime mover of economic growth. In another paper in the same year, Nelson states: "It is basic research, not applied research, from which significant advances have usually resulted" (Nelson, 1959b: 301). "It has become almost trite to argue that we are not spending as much on basic research as we should. But, though dollar figures have been suggested, they have not been based on economic analysis . . . How much should we spend on basic research? Replacing the X of the familiar literature on welfare economics with 'basic research' provides the theoretical answer" (Nelson, 1959b: 297). Nelson contrasts the private costs of research activities to the "obvious" social returns of research to make a case for more basic research – social returns became a much-studied topic later on (Griliches, 1958; Mansfield et al., 1977; Minasian, 1969).

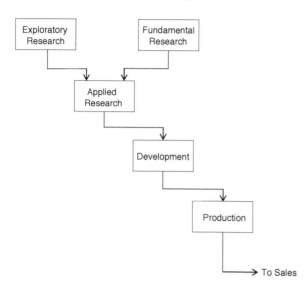

Figure 1.1: *Clifford Furnas's flow diagram from research to sales*

Note

1. For example, in 1960 the US President's Commission report on "Goals for Americans" included a chapter on technological change (US President's Commission on National Goals, 1960). The President's Commission was followed by the creation of a National Commission on Technology, Automation and Economic Progress in 1964 (US National Commission on Technology, Automation and Economic Progress, 1966). In 1964, the Bureau of Labor Statistics produced a voluminous study on "Technological Trends in 36 Major Industries" for the President's Advisory Committee on Labor-Management Policy, updated in 1966 (US Bureau of Labor Statistics, 1966). The same year, the US Council of Economic Advisers had a chapter titled "The Promise and Problems of Technological Change" in its annual president's report (US Council of Economic Advisers, 1964).

PART I

TECHNOLOGICAL INNOVATION

The term technological innovation is composed of two concepts: technology and innovation. According to historian Eric Schatzberg, technology has had two meanings or uses over time: industrial techniques, and knowledge pertaining to these techniques (Schatzberg, 2018). A third meaning needs to be added to Schatzberg's definitions. After the Second World War, technology came to mean goods for the market, goods that embody knowledge. The innovation of technological innovation helps to make sense of this meaning. The concept has two key meanings. One common use is as a synonym for novelty (*novus*) or invention. The other use is as action. As the etymology suggests (*in+*), innovation is introducing a novelty in practice – implementing, applying, adopting, putting into use. In the case of technological innovation, *in* refers to the commercialization of new goods.

- *Introduction:* introducing something new to the world. This concept first appeared among anthropologists and sociologists, but is most popular among economists and management.

- *Application:* assimilation, transformation, exploitation, translation, implementation: applying (new) knowledge in a practical context. Innovation is the application of ideas, inventions and science.
- *Adoption:* acceptance, utilization, diffusion, transfer: adopting a new behavior or practice. These concepts are mainly used by sociologists.
- *Commercialization:* bringing new goods to the market. Used concurrently with introduction or application, this concept applies to industrial innovation.

Another difference is between technological innovation and technological change. Technological change refers to changes in productivity due to changes in factors of production or industrial techniques. Technological change is concerned with the use of new industrial techniques in production, called processes, and their effects on productivity. The other term (technological innovation) refers to the commercialization of new goods. Technological innovation includes both industrial processes and products or goods for consumption, but concentrates on the latter. Industrial processes are considered to the extent that they are goods for the market (firms buy processes). Technological change as a research tradition is often overshadowed by the research tradition on technological innovation, although the latter constantly uses the results coming from the former as a source of empirical evidence for the contribution of R&D to productivity.

The context out of which technological innovation as a term emerged is threefold. First, economic growth.[1] After the Second World War, the primary economic issue was no longer full employment but growth (Arndt, 1978). There was a fear of stagnation, and hence programs of reconstruction and productivity. In this context, technological innovation came to be considered a source of national growth (rather than a source of unemployment). In his presidential address, delivered at the 51st annual meeting of the American Economic Association in Detroit, Michigan, 28 December 1938, Keynesian economist Alvin Hansen claimed that business recoveries were weak and depression prolonged and deep, causing "secular stagnation". He argued the real problem of the time was "the character of technological innovations", namely the absence of large investments of capital (Hansen, 1939: 4, 5). Hansen distinguished two types of investments: deepening of capital (more capital used per unit of output, as a result of technological change I would say) and widening capital (real capital growth). To Hansen, the latter did not increase sufficiently, partly because of a declining rate of population – population being potential customers for new goods. "In the beginning stages of modern capitalism both the deepening and the widening processes of capital formation were developing side by side. But in the later stages, the deepening process, taking the economy as a whole, rapidly diminished. And now with the rapid cessation of population growth, even the widening process may slow down" (p. 7). "A full-fledged recovery ... requires a large outlay on new investment ... But such new developments are not currently available in adequate volume" (p. 11).

In this context, to many economists, science became synonymous with economic growth. "A very large share, if not the bulk, of the increase in output is to be attributed to advances in knowledge", stated economist Moses Abramovitz, in an early review of the emerging literature on economic growth (Abramovitz, 1952: 141). At about the same time, economists Walter Rostow (1952), Simon Kuznets (1959) and François Perroux (1961) discussed "innovation" as a factor explaining economic growth. Rostow talks of "propensities" determining economic growth, Kuznets and Perroux of "conditions". To take just one example, to Kuznets the "necessary condition" for economic growth is innovation, defined as the "application of new bodies of tested knowledge to the processes of economic production ... Science is the base of modern technology, and modern technology is in turn the base of modern economic growth" (Kuznets, 1959: 14–15, 30). "The epochal innovation that distinguishes the modern economic epoch is the extended application of science to problems of economic production" (Kuznets, 1966: 9).[2] By growth, Kuznets means "a sustained increase in the output of goods" or an "increase in per capita product (Kuznets, 1959: 13–14), arising from a shift in relative proportions of various goods demanded and used – and hence major changes in combinations of productive factors" (p. 14).

The development of the System of National Accounts (United Nations, 1953; OECD, 1958)[3] and growth accounting[4] provided the opportunity to governments to set targets for growth. In 1961, the OECD suggested a target of 50 per cent growth in gross domestic product (GDP) for the decade 1961–70. The United Nations followed with its own targets: 5 per cent for the decade 1961–70 and 6 per cent for the next decade. The scientific community, broadly defined, developed its own accounting and targets of attainment, defined in terms of gross expenditures on R&D (GERD) as a percentage of GDP (National Science Foundation, 1959; OECD, 1963c). In light of "disparities" in R&D expenditures between countries, or "gaps" as they were called, the US GERD/GDP ratio of the time (3 per cent) was idealized as the standard norm (Godin, 2005).

A second element of context, closely linked to economic growth, is international competitiveness (trade). In the early natural philosophers' theory of stages of development, some countries were declared more advanced than others. There are advanced and backward countries. The theory lately got converted into stages of economic development, starting with Karl Marx's *Das Capital* of 1887, then stages of economic growth

(Rostow, 1960). Theories of innovation followed. Innovation and the diffusion of innovation in time and space is explained as a "sequential spread" of pioneers and followers. As Simon Kuznets put it, major innovations "originated in one part of the world, and then spread first to the major followers [United States, Germany, Britain], then to the next set of countries [Japan, USSR]" (Kuznets, 1959: 114–15). Trade theory supported these representations: innovation follows a life cycle. Innovations emerge in advanced countries, then diffuse to other countries as they mature (Posner, 1961; Vernon, 1966). At the heart of this discourse is the idea of the "time lag" between invention and commercialization – a discourse first proposed by Maurice Holland from the National Research Council (Holland, 1928) – lags that firms and countries manage with varying success. National performances in productivity and competitiveness are explained by performance in innovation. There are leader and laggard countries. This led to the notion of "technological gaps", a much-discussed topic in the 1960s.

National concerns surrounding growth and competitiveness were just two elements of the attention paid to technological innovation. A third was organizations, or the "organizational society" as some called it (Boulding, 1953; Parsons, 1956; Presthus, 1962). "Our economy can be conceived of as a matrix of several million highly interrelated organizations ... Organizations are the major mechanisms for achieving man's goals"

(Hage and Aiken, 1970: 5). It is through firms that national growth occurs. In a context of change (or perceived change), organizations were urged to innovate. It is a question of "survival" for organizations to respond to changes in the environment, namely to new technologies and new markets.[5] A huge literature soon developed to discuss organizational innovation and the management of technological innovation.[6]

In spite of this context, it is above all, from considerations or interests in research, as a factor of economic growth that the discourse on technological innovation has emerged. Research is the source of innovation, so it was said. As American philosopher Donald Schon, author of the critical book on innovation, *Technology and Change* (1967), and successively consultant at Arthur D. Little Inc. and Director at the Department of Commerce, then professor at MIT from 1968 until his death, put it: "Research came to be seen as the instrument for growth, and growth as the occasion for and object of research" (Schon, 1967: 54).

Notes

1. The literature on economic growth is voluminous, particularly from the late 1930s (see Roy Harrod, Evsey Domar, Robert Solow, Nicholas Kaldor, Joan Robinson, Paul Romer and Robert Lucas). For general and historical overviews of economic growth (and development) theories, see Arndt (1978, 1981, 1987), Hicks (1966) and Rostow (1990).
2. Like Joseph Schumpeter, Kuznets concentrates on "major innovations", what he later calls "epochal inno-

vations" (Kuznets, 1966). He distinguishes discovery, invention, innovation and improvement. Discovery (an addition to knowledge), invention (a tested combination of already existing knowledge to a useful end), innovation (an initial and significant application of an invention, whether technological or social, to economic production), improvement (a minor beneficial change in a known invention or process in the course of its application) (pp. 30–31).

3. For some history, see Carson, 1975; Coyle, 2014; Masoon, 2016.

4. For example, National Bureau of Economic Research; Robert Solow, Moses Abramovitz, Salomon Fabricant, John Kendrick, Edward Denison.

5. One manager suggested that the organization's motive to innovate is not the "ancient" economists' assumption of profit maximization (not only), but the survival to external threats (Gellman, 1970: 130).

6. For example: Argyris, 1965; Becker, 1964, 1967; Burns and Stalker, 1961; Carter and Williams, 1957, 1958, 1959; Hage and Aiken, 1970; Lorsch, 1965; Twiss, 1974.

2. Innovation as science applied

The private firm as a basic form of organization of society is the concern that originally gave rise to thoughts on technological innovation. In order to survive and compete, the firm must innovate, namely by developing and using science. The imperative made use of the arguments of scientists and their representative organizations as to the effect of science on economic progress. In turn, scientific organizations and public agencies commissioned studies to document the case.

The British Association for the Advancement of Science

In 1952, the Science and Industry Committee of the British Association for the Advancement of Science (BAAS), an organization founded in 1831, contracted a study on "the problems of speeding up in industry the application of the results of scientific research" to two of its members: Charles Carter (Chairman) from Queen's University of Belfast and Bruce Williams (Secretary) from University College of North Staffordshire. The background to the study was the debate on time lags in the application of science in industry. Britain is slow to apply the results of science, it was said. This is an old and recurrent issue in Britain (Henderson, 1927; Manchester Joint Research Council, 1954). The BAAS study was supported by the Society of Arts, the Nuffield Foundation (for social wellbeing), with funds from the Department of Scientific and Industrial Research (DSIR) and the Board of Trade (funds coming from the grant from US economic aid). The study was conducted between 1952 and 1956, and led to a series of three books: *Industry and Technical Progress* (1957), *Investment in Innovation* (1958) and *Science in Industry* (1959). These are certainly among the first books ever written on industry and technological innovation, understood as "science applied", in the sense of the use or application of science in industry (a variant of or a declination on the concept of applied science).

I talk here of science applied rather than applied science because, first, applied science is only one step or stage in the "circuit" of the innovation process (that consists of basic research, applied research, development). Carter and Williams call this process "technical progress".[1] It "involves the movement of an idea from its first discernible beginnings, often far from any possibility of application, to its successful commercial use" (Carter and Williams, 1957: 4). This is what came to be called the "linear model of innovation", a model first imagined by Rupert Maclaurin in the 1940s. Second, again and again, Carter and

Williams stress that "technical progress does not always require the application of new *scientific* knowledge".[2] It may build on existing knowledge as well, which is "adapted, developed, tried out in new circumstances" (Carter and Williams, 1959a: 3).

Carter and Williams's *Industry and Technical Progress* (1957) is based on a survey of 246 firms and 109 universities, public agencies and departments, and research and trade associations. It studies the factors and conditions of technological innovation according to the "circuit" or "stages" of application of knowledge in industry, from basic research to development to production. The factors studied as leading to technological innovation are: personnel (scientists and engineers), size of organization, management, financial resources, market and environment. To Carter and Williams, eliminating the "gap" between invention and innovation is a matter of "industrial growth", a matter also of "survival" in a competitive world. Innovation is the latest but most important stage of an "investment decision", namely "the action of bringing the new idea into practical use" (Carter and Williams, 1957: 76). [3]

Investment in Innovation (1958), a subsidiary report to *Industry and Technical Progress*, concentrates on firms' "investment decision" to innovate as a "desire to survive" in a competitive market. The report studies 204 product, process and method innovations, looks at their sources, and offers a classification of innovations as "passive" (responding to a market pressure) or "active" (deliberate searching for new markets). Such a classification led to many others in the following decade – such as the OECD study on *Gaps in Technology* (1968–70) and economist Chris Freeman's categorization of firm strategies as offensive, defensive, imitative, dependent, traditional and opportunist (Freeman, 1974).

The third study, *Science in Industry* (1959), extends the previous two analyses to public policy. "We doubt if it can be said that a Government policy on the application of science really exists . . . The present pattern of Government intervention in favour of technical progress has been formed haphazard from a number of pieces shaped by past history" (Carter and Williams, 1959a: 103). The existing aid is dispersed in many departments but "a general 'Government scientific policy' is something more than the mere addition of the policies of separate Departments" (p. 100). What is an innovation policy or "strategy"? According to Carter and Williams, innovation policy is concerned with a large range of government measures, such as public support to research and development, education, taxation, competition and trade.[4] This is perhaps the first consideration of policy in discussions about technological innovation. As discussed below, such a diversity of measures or packages – a "policy mix" as some call it – would come to define innovation policy in the following decades.

Carter and Williams's studies are quite original

and were influential. They paved the way for the study of technological innovation, a kind of study imitated by many others in the following decades. The two dimensions of innovation studied – firm and policy – gave rise to two key features of subsequent discourses on innovation. The consideration of policy in particular is quite unique for the time. Carter and Williams considered innovation policy in broad terms, as would be the case in the next decades. The focus on a time lag and gap in the use of science in industry also gave rise to a passionate public debate on technological gaps in the 1960s, and Carter and Williams's taxonomy on technical progressiveness later led to a much-cherished indicator: high technology.[5]

The Department of Scientific and Industrial Research

When Carter and Williams were beginning to communicate the results of their study (see Williams, 1956), the Department of Scientific and Industrial Research (DSIR), created in 1916, commissioned its own study of industrial innovation from Tom Burns, Professor of Sociology, and George Stalker, Professor of Psychology, both from Edinburgh University. The study was an extension to the whole of Britain of a previous study of 20 Scottish firms in the electronics industry, a study funded by the Scottish Council and conducted by the same authors between 1953 and 1957.[6]

The British extension to the study was conducted between 1957 and 1960. The result is a monograph titled *The Management of Innovation* (1961). To Burns and Stalker (1961: 33):

> The reunion of industry and science through the new technology effects a radically different, much more intimate combination of forces than anything that has obtained before (p. xxxv). Technological progress has become of vital concern for the individual firm in many industries, and the increasing pace of innovation makes it inevitable that the firm provide more and more support for research and development as a condition of its own survival.

To the authors, innovation, not defined explicitly in the book (a common practice in the following decades), refers to new products and industrial processes.[7] The purpose of the authors is to understand "The connexion between progress in material technology and the emergence of new forms of social organization" (Burns and Stalker, 1961: 19). "How management systems changed in accordance with changes in the technical [technological advance] and commercial tasks [mass production] of the firm, especially the substantial changes in the rate of technical advance" (p. 4). The author offers two ideal types of organization, one adapted to a bureaucratic environment, and the other to a changing organizational environment, the first mechanistic (a formal organization, bureaucratic, with programmed decision making) and the second organic (an informal organization with non-programmed decision making), the former being appropriate for stable conditions and the latter required for conditions of change (pp. 119–22). In the face of change in the environment, the challenge of an organization, on which its survival depends, is to shift from a mechanistic to an organic mode of "management system". Burns and Stalker stress that this is not a dichotomy, but rather a polarity whose extremes are stability and change (p. 122).

Burns and Stalker have had a certain impact among a few scholars who revisited the typology of organization forms in the 1970s. Above all, *The Management of Innovation* inaugurated the study of organizational innovation. It also frames the study of technological innovation in terms that would become conventional in later decades: a matter of two factors, supply (technology) and demand (market).

The representation of innovation as the application or "exploitation of research" in industry is one of the first representations of technological innovation. Many scholars of the time espoused the representation – with studies of the phenomenon, it must be said, varying in depth. Innovation is:

- The application of new discoveries on a commercial scale (Boulding, 1946: 86).
- Practical applications of fundamental and applied science (Rostow, 1952).
- [The use of new technical knowledge] in order to derive new products (Burns and Stalker, 1955: 249).
- The application of new knowledge (Carter and Williams, 1958: vii).

- A new application of either old or new knowledge to production processes (Kuznets, 1959: 29).

The representation stresses an activity. Certainly, this activity is conducted with an ultimate outcome in view. Yet the outcome is not called an innovation. Today, we regularly use innovation in the sense of outcome. Scholars do too. Innovation is a semantically fluid concept.

Notes

1. To the authors, technical progress and innovation are synonymous. However, the former has a far larger place than innovation in the books.
2. At the same time, the authors state elsewhere that innovation is the "application of new knowledge" (Carter and Williams, 1959a: vii).
3. The authors also use the concept of innovation in the sense of outcome.
4. Carter and Williams suggest two principles for policy: aid to progressive industries but, most importantly, aid to promising but backward industries, among them the small- and medium-sized enterprises.
5. In *Industry and Technical Progress*, the authors developed a taxonomy of "technical progressiveness", based on 24 characteristics (see also Carter and Williams, 1959a).
6. For a summary, see Croome, 1960a, 1960b. Beginning in the early 1950s, Burns had conducted a series of studies on how firms cope with and adapt to a changing environment.
7. For explicit definitions, see Burns and Stalker (1955: 249): "adoption" of new technical knowledge "in order to derive new products"; Burns (1956: 147, fn1): "an invention becomes an innovation when it is carried into commercial application".

3. Innovation as outcome

By the early 1960s, innovation was the word on everyone's lips. In February 1960, in a talk presented at a joint meeting of regional groups commemorating National Engineers Week in Schenectady, New York, Herbert Hollomon, an engineer and Head of General Electric Engineering Laboratory, and soon to become Secretary of Commerce, embraced innovation to support his idea for a National Academy of Engineering, which was in fact created in 1964. In his view, innovation is the job of the engineer (Hollomon, 1960). In the same year, John Gardner, President of the Carnegie Foundation, and Secretary of Health, Education and Welfare under President Lyndon Johnson, wrote a chapter on education in the president's report, *Goals for Americans*. This report included chapters on science (by Warren Weaver, scientist and administrator) and technological change (by Thomas Watson, President IBM). Like Hollomon in his call for innovation, Gardner urged for "a more vigorous tradition of research in education" or educational innovation (school organization, teaching aids) (Gardner, 1960: 88). A few years later, Gardner produced a book on "innovative society" and the imperative for continuous creativity and innovation. "The pressing need today

is to educate for an accelerating rate of change",
Gardner stated (1963: 22).

Both the BAAS's and the DSIR's commissioned
studies (see Chapter 2) were concerned with the
industrial or firm level. The next series of studies
extended the argument to the national level. At
the US National Science Foundation (NSF), an
organization created in 1950 to fund basic research,
thoughts on technological innovation emerged
out of a discourse on research, and as a legitimiza-
tion of research expenditures. The NSF was inter-
ested in technological innovation to the extent that
it was an outcome, arising from basic research.
Over the following years, the NSF commissioned
many studies to determine to what extent this was
true, and to determine what links existed between
science and economic growth.

In 1953 the NSF started conducting regular
surveys of research and development in the
country. The objective was, as stated by-laws of
the organization, to assess the value of funding
basic research (in the face of skeptics). At the
same time, the surveys served the NSF discourse
on research funding needs. A few years after the
surveys began, Raymond Ewell, a chemist and
Vice-Chancellor for Research at the University
of Buffalo, on loan to the NSF since 1953, conducted a statistical analysis on the relationship
between research and economic growth. At the
NSF, Ewell's major task was establishing a study
group to examine the economic effects of research.
The study was expected to last five years. It took

longer to materialize. Results only began to appear in the early 1960s.

In the meantime, Ewell produced what was perhaps the first analysis of its kind, presented at the annual meeting of the American Drug Manufacturers Association in April 1955. To Ewell, "Research may be the most important single factor in economic growth in the United States . . . Research is the spearhead of economic growth . . . [It] is a highly creative activity – it produces new products, creates new jobs and new industries, cuts costs of production, and makes a large contribution to our economic growth and our over-all national welfare . . . A steady flow of technological advances is the best protection against business depressions" (Ewell, 1955: 2980).

Ewell's demonstration method was quite simple. Making use of the newly conducted surveys of research and development at the NSF, and fragmentary data going back to the 1920s, Ewell correlated research and development expenditures with gross national product (GNP) growth, a very familiar method in the following decades.[1] Ewell calculated that the growth rate of GNP has been 3 per cent per year since 1910, while research and development grew by 10 per cent. This shows "a high degree of correlation" (Ewell, 1955: 2981–2). "The GNP would have been only $285 to $325 billion in 1953 if no research had been conducted since 1928", representing cumulatively a loss of GNP of US$400 to $800 billion between 1928 and 1953 (p. 2984). This amounts to an annual return on research investment of 2,700 to 5,400 per cent. The amount spent on research and development in Ewell's time ($4 billion/year) suggests a cumulative increment to GNP of $100 to $200 billion in another 25 years. "We should probably be putting more of the national income, or of the national effort, into research than we are now doing" (p. 2985).

With so broad a range of estimated values, Ewell admitted the results to be "educated guesses", with "a margin of error of a factor of two" (p. 2983). Yet the contribution of research to the economy for 1928 to 1953 was illustrated with numbers on specific contributions, among them the introduction of new products ($5–10 billion), lowering costs of production (improved industrial processes) ($10–20 billion) and expanding markets ($5–10 billion).

The NSF extended the methodology in subsequent years. In 1961, the organization published the study "Research and Development and the Gross National Product". Instead of statistically correlating research and development expenditures with GNP, as Ewell had, the organization examined the volume of research and development expenditures in terms of, or in the "framework" of, the National Accounts. Over the years, the NSF used a matrix, first developed at the Department of Defense, then used at the NSF after 1956, that cross-references performers of research and development according to the source of funding (Godin, 2005) – Ewell may have been the source of the idea. The NSF study demonstrated that research and development expenditures amount to 2.6 per cent of GNP in 1959, the majority of activities conducted or performed in the private sector. However, the government appeared as a major funder, being responsible for almost half of the private sector.

In this study, the NSF made use of a new term: technical innovation. Research is an "investment" toward technical innovation. It has a "multiplier effect which yields a return in the future many times the initial outlay" (US National Science Foundation, 1961: 5). The study's message was the same as Ewell's, with a new term: "Since technical innovation derives ultimately from basic research, the small part of national resources currently represented by this scientific effort suggests that additional investment in this area is a most promising and relatively inexpensive means of promoting long-term economic growth" (NSF, 1961: 5). Yet "too little is presently known about the complex of events to ascribe a specified increase in gross national product directly associated with a given research and development expenditure. Various studies are needed – some of which are under way by the Foundation" (p. 7).

A conference on "Research and Development and its Impact on the Economy" dated 20 May 1958 announced a series of further studies. The conference gathered 400 participants. "It was the consensus of the Conference that more funds should be channeled into basic research" (NSF,

1958a: 3), a message that Alan Waterman, Director at NSF, disseminated year after year to members of the US Congress.[2] Waterman announced that the organization would engage in special studies to determine the impact of research and development on economic growth. The first study, commissioned from the Case Institute of Technology, was concerned with "developing a methodology and 'model' for determining the relation of research and development to the growth of the company". The second study, on "the effect of decision-making as it relates to innovations resulting from research and development", was commissioned from the Carnegie Institute of Technology (NSF, 1958b: 155).

The first series of studies appears to have produced no output.[3] More fruitful was the series that the NSF Office of Special Studies commissioned from the young economist Edwin Mansfield. Originally published in the NSF series *Review of Data on Research & Development* in 1961 and 1962,[4] the studies looked at the diffusion or "first introduction" (commercial application) of 14 major inventions (called innovations) (Mansfield, 1961, 1962).[5] Mansfield transformed these studies into scholarly articles and built a career on these early works. In the following decades, he published a great number of frequently cited articles and books on technological change and technological innovation.

This was only a beginning. The NSF continued to commission studies on technological innovation in the following years. In December 1962, it requested a study on "technical innovation" from consulting firm Arthur D. Little Inc.: the sources, problems and patterns of technological innovation and "the process by which inventions and discovery occur and are used" (Arthur D. Little Inc., 1963). The consulting firm was certainly the ideal candidate for such a study, having conducted one on science and industry for the National Research Council and the National Resources Planning Board in 1941 (US National Resources Planning Board, 1941). It studied "significant" technological innovations in five industries (significant in terms of economic and technical importance and the degree of change required in the organization), innovation being defined as one stage in a continuum (invention, innovation, diffusion): "the process of bringing invention into commercial use" (Arthur D. Little Inc., 1963: 6). Two main results emerged from the study. First, technological innovations come from "science-based" industries (those industries that invest in research and development). Second, most major technological innovations come from outside the firm that uses them (innovation by invasion).

The report gave rise to a very interesting book, perhaps one of the best books on technological innovation for a while, from Donald Schon, philosopher and consultant at Arthur D. Little Inc., and responsible for examining the semiconductor and appliance industries in the first study: *Technology and Change: The New Heraclitus*.[6] Schon reanalysed the material from the 1963 study, and produced a critical analysis of the then-current discourse on technological innovation, defining innovation as a "process of bringing invention into use" (Schon, 1967: 1). Schon contrasted the "rational" view of technological innovation, from basic research to production, a process that can be managed and controlled (linear; goal directed; and intellectual) – a view of technological innovation "as we would like it to be" (p. 37) – to a non-rational process (social; working backward, with feedback, accidents and uncertainties; driven by market needs rather than research).

In the meantime, the NSF started work with the National Planning Association, a private non-profit organization founded in 1934, from which it commissioned a study to explore the factors involved in the civil application of government-sponsored research and development, namely "the source and impact of externally generated science that have served as a stimulus to technological innovation". The study was conducted between 1963 and 1967, under the direction of Sumner Myers, who had a bachelor's degree in mechanical engineering and a master's from the MIT Sloan School,[7] with James Utterback as assistant (and subsequent promoter of the framework). Preliminary reports appeared in 1965 and 1967 (Myers, 1965, 1967; Myers and Sweezy, 1965). Subsequently, psychologist Donald Marquis, formerly of the Department of Defense and then Director of the Management of Technology program at Sloan

School of Management (MIT), joined Myers for the final report *Successful Industrial Innovations* (Myers and Marquis, 1969).[8]

Successful Industrial Innovations is a statistical study of 567 "innovations" introduced by 121 firms in five manufacturing industries. The concept of innovation is used here as a substantive (novelty). But the report stresses the transitive form. Defining technical innovation as a process – "a complex activity which proceeds from the conception of a new idea to a solution of the problem and then to the actual utilization of a new item of economic or social value . . . a total process of interrelated sub processes" (Myers and Marquis, 1969: 1), rather than as a single action, and one that does not always occur in a linear sequence[9] – the study developed a framework influential for years to come. Based on a classification and analysis of the sources of information in the innovation process, the report stresses that technical innovation depends not only on research (supply), but on both technical feasibility and (potential market) demand or needs, as Tom Burns and George Stalker stated before, fused into a "creative act" (see Figure 3.1).

Need or demand was part of a leading discourse in the late 1960s, and the NSF took part in a controversial debate. Between 1963 and 1969, the Department of Defense conducted a study to identify "management factors for [productive or useful] research and technology programs", and to measure "cost-effectiveness" or return on the department's investment in research.

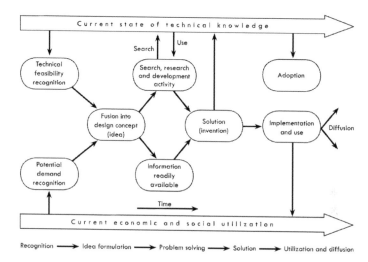

Figure 3.1: Sumner Myers and Donald Marquis's
Model of the Innovation Process (1969)

Project Hindsight, as it was called, examined 20 weapons systems and other military equipment, and traced the post-Second World War contributions of research and development (called Events) backward.[10] The results were published between 1967 and 1969. The project determined that most weapons systems rely on research of the applied type, rather than basic research: "Nearly 95 per cent [of innovations in weapons systems] were motivated by a recognized Defense *need*", stated the report on Project Hindsight. "Only 0.3 per cent came from undirected science" (Sherwin and Isenson, 1969: 1577; US Department of Defense, 1969).

The NSF contested the Department's results using its own numbers (*project TRACES*). The contractor (Illinois Institute of Technology) studied "five major technological innovations" (in terms of socioeconomic importance)[11] and, using the same vocabulary as the Department of Defense (events), found that the majority of key events leading to the innovation depended on non-mission research, namely research motivated by the search for knowledge without any regard to its application (Illinois Institute of Technology, 1968). This was a far cry from the results of the Myers study. A long-lasting controversy followed among scholars as to whether technological innovation is science-push or market-pull or both. In the end, most scholars of the time agreed that both factors are "coupled" together in the process of technological innovation (see Langrish et al., 1972; Rothwell and Robertson, 1973; SPRU, 1972). Research plays no exclusive role – "innovation-oriented research" perhaps being the exception (OECD, 1971: 103).

Notes

1. Previous accounting exercises calculated a "national science budget" and estimated its share in the national income (with comparisons between countries), the sources of funding by economic sector and the types of research conducted (Godin, 2005). None correlated research and development with GDP, together with estimates of economic growth.
2. For example, in testimony to the Joint Economic Committee of the Congress chaired by Paul Douglas in February 1959. Waterman recalls the 1959 *Economic Report of the President* that "called attention to the

extremely important role that research and development contribute to the growth of the economy" and
to the Committee for Economic Development's report
Economic Growth in the United States (produced by
Edward Denison in 1962) that gave "prominence to
the force of 'technological change' and advancement".
Denison stressed the "overwhelming evidence of
history that new labor saving devices ultimately create
new industries, new investment opportunities" (NSF,
1958a).

3. The Case Institute of Technology (1947–67) emerged
out of the Case School of Applied Research (Case
Western Reserve, Cleveland, Ohio University). Russell
Ackoff and Burton Dean were active researchers in
management and operations research. Of the two, Dean
produced works on research and development, but
nothing that resembled anything like an output from
the NSF's original commission. A third researcher was
Bela Gold, who produced a series of papers and books
on industrial innovation, but again, nothing that points
to an output from a commissioned NSF study. A fourth
professor was Herbert Shepard, who specialized in
organizational development. Finally, Melin Krazberg
was a historian of technology, and was not involved in
statistical work.

4. A summary appeared in the issue of March 1963.

5. Mansfield produced another study for the organization in 1972 (later published as Mansfield, 1972). At
the request of the Office of Management and Budget, a
department that has always been skeptical of research
expenditures, the NSF organized a colloquium in
April 1971 on research and economic growth, based
on a series of review papers commissioned from four
economists: Charles Stewart (George Washington
University), Edwin Mansfield, Zvi Griliches and
William Fellner (Yale University) (NSF, 1972). The
colloquium discussed the many problems and limitations in measuring the relationship between research,
growth and productivity.

6. Schon worked at Arthur D. Little Inc. from 1957 to 1963, before he joined the Department of Commerce from 1963 to 1966. As Director of the Office of Technical Services, Department of Commerce, Schon was member of an OECD ad hoc group on the development and exploitation of invention and research results (1963–4).

7. Myers worked in private industry before joining the Institute of Public Administration, specializing in urban planning and transportation.

8. The collaboration may be due to a convergence of interests on technology transfer. In the course of the study, Myers organized a conference on technology transfer and innovation in the spring of 1966 (NSF, 1966). In May of the same year, the MIT Center for Space Research organized a conference on the same subject (Gruber and Marquis, 1969).

9. Innovation is also "defined broadly as the introduction by a firm of a technical change in product or process ... the first use of a technical change which is new to the economy as well as new to the firm", namely original as opposed to imitation or adoption (Myers and Marquis, 1969: 3).

10. "An RXD Event is a period of technical activity with a well defined outcome ['progress report, proposal, journal article, patent disclosure or some other document'] which has influenced the development of weapon systems" (Little, 1965: I–1).

11. Contraceptive pill, matrix isolation, video tape recorder, magnetic ferrites, electron microscope. A subsequent study added five case studies: heart pacemaker, hybrid grains, electrophotography, input–output economic analysis and organophosphorus insecticides (Battelle-Columbus Laboratories, 1973).

4. Innovation as process

By the late 1960s, innovation had become a buzzword. As Jack Morton, the engineer at Bell Laboratories who brought the transistor from invention to market, and author of numerous articles and a book on technological innovation, put it in 1971: "Innovation is certainly a 'buzzword' today. Everyone likes the idea; everyone is trying to 'innovate' and everyone wants to do better at it tomorrow" (Morton, 1971: 73). Because of, or thanks to, "technological innovation", the word innovation has become part of our everyday vocabulary. Morton himself was part of this process. From the mid-1960s onward, he developed thoughts on technological innovation that culminated in the publication of *Organizing for Innovation* (1971) (more on Morton in Chapter 5).

Two concepts serve to make sense of and rationalize the discourse on technological innovation: process and system. Both give place to research, but the ultimate phase is commercialization, whose key agent is the firm. Over time, the discourse on research gave (almost) full place to a discourse on innovation. A new representation of technological innovation developed that placed emphasis on the final stage of the sequential process (commercialization) rather than on the first stage (research). In that representation, technological innovation refers to (1) a good (output) for the market, a good that embodies science, or knowledge or research certainly, but above all, (2) its commercialization. The semantic shift here is from a concern with universities to a concern with firms, from research to marketing. A key feature of the discourse is technological innovation as the "first" commercial production of an invention.

Process is a concept that has a long history (Godin, 2017). Technological innovation is a process in time, a series of activities whose ultimate purpose is technological innovation, as Sumner Myers put it. This is the subject of the present chapter. Technological innovation is also a process, in the sense that it is a system, composed of diverse institutions, all acting toward a shared purpose: the use and diffusion of technological invention in practice. The system dimension of technological innovation will be addressed in Chapter 5.

The OECD

Back to back between 1966 and 1968, one international organization and two government agencies produced policy documents on technological innovation, the first such documents ever produced. The first organization was the OECD.

Research and economic growth

Like the NSF, the OECD's work on technological innovation is a spin-off from reflections on basic research. In 1963, the newly created organization (1961) and its Directorate for Scientific Affairs organized a ministerial conference on the contribution of research and development to the "modern 'categorical imperative' – economic growth" (OECD, 1963b: 53). The background document to the conference, *Science, Economic Growth and Government Policy*, was written by Chris Freeman (National Institute for Economic and Social Research, London), Raymond Poignant (Délégation générale à la recherche scientifique et technique, Paris) and Ingvar Svennilson (Institute of Social Sciences, Stockholm). John Cockcroft (Churchill College, Cambridge, and member of the British Advisory Council on Scientific Policy) and Michael Michaelis (Arthur D. Little Inc., consultant to the US Office for Science and Technology, and member of the National Planning Association's Advisory Committee on Science, Technology and the Economy, which supervised the Myers study) "helped" with preparation of the report.

The word innovation is everywhere in the OECD document, understood broadly as "constantly introducing new products and processes" (OECD, 1963b: 30) and narrowly as a commodity arising out of research (output).[1] However, the focus of the report is research, as an "investment in innovation" – a key phrase in the report.[2] "If the targets for economic growth set by the Ministerial Council of the OECD are to be reflected in policy, it is evident that scientific research and technological development will form one of its main instruments . . . The collective target for economic growth set by the OECD for its Member countries, if it is to have its full meaning, should be paralleled by co-ordinated policies for science and technology" (pp. 10–11).

The report documents "how research and development activities could be made to bear in the most efficient way on economic growth" (p. 11). Education and research, or "technical progress", claim the authors, account for 90 per cent of the increase in GDP (the report cites Denison, 1962; Massell, 1960; Solow, 1957). But on average, countries spend only 1 per cent of GNP on research and development, compared to 3 per cent in the United States.

To raise their ratio, governments should support industries, particularly "research-intensive industries". But instead, research policy is uncoordinated. There exists a multiplicity of motives rather than a "national strategy" as Charles Carter and Bruce Williams claimed some years before. "There is practically no such thing as a science policy, but rather a series of separate science policies, aimed at different objectives" (OECD, 1963b: 50). Government policy should aim at a "comprehensive science policy" (p. 59) that includes direct financing, indirect support (tax policy, contracts, cooperative research, communication and information services, education, balance between fundamental – free and "orientated" – and applied research) (pp. 63–4), collaboration between government bodies responsible for military and civil research, and international cooperation. There is also a need for "a dialogue between those responsible for economic policy and those responsible for science policy" (p. 69).

This message got into the pages of *Nature*. In October 1963, *Nature* published a talk given before the British Association for the Advancement of Science meeting in Aberdeen the previous September, titled "Research, Innovation and Economic Growth". The author was Keith Pavitt, from the OECD Directorate for Scientific Affairs (Pavitt, 1963). The article was a spin-off from the background document to the first OECD ministerial meeting. Yet, Pavitt gave a twist to the document and put it entirely in the light of technological innovation. In the document, Pavitt discusses the role of research and innovation in economic growth, and notes the need for policies that "take due account of their possible impact on the economy". He surveys diverse policies[3] and offers a threefold typology of policies to stimulate innovation:

- Science-intensive industries
    - Development contracts
    - Government buying procedures
    - Spin-off from military and space research
    - Grants for developing inventions

- Old-established industries
  - Cooperative research
  - Information and advisory services
- Other measures
  - Research institutes
  - University–industry cooperation
  - Fiscal measures

In line with economist Simon Kuznets, Pavitt defined innovation as a "new application of either new or old knowledge to production processes". To Pavitt, innovation depends largely on research and development, and fundamental research is in the long run an "essential pre-requisite" – a message quite different from that of later OECD reports from 1966 onward (see below). Briefly stated, the ideas of a process composed of several steps, or a system composed of several active participants, and the contribution of many factors, were not yet considered, although Pavitt does discuss a broad range of policies, from education to research to procurement to fiscal measures.

Government and technical innovation

From its early beginning, the OECD, through its Directorate for Scientific Affairs, was concerned with research and "science policy" (see OECD, 1963a). Just a few years after the first ministerial meeting on science, the organization added innovation to its agenda. As a follow-up to the 1963 meeting, a working party was created in January

1964 to study the "means available to governments for stimulating research in industry and for accelerating the pace of innovation in the civil economy" (OECD, 1964: 1). Innovation is here understood to mean the "effective use of scientific knowledge in the economy". To support the second ministerial meeting (1966), the OECD produced one of the first titles on technological (technical) innovation ever: *Government and Technical Innovation*, written by Keith Pavitt.

To the OECD, the principal benefit of technical innovation is economic growth and competitiveness. "There is little doubt that if governments succeed in helping to increase the pace of technical innovation, [this] will facilitate structural changes in the economy, and increase the supply of new and improved products necessary for Member Governments to achieve rapid economic growth and full employment and without inflation" (OECD, 1966: 8). The responsibility for achieving this result basically rested on firms. "The competitive position of a firm now depends ... on the speed with which it can introduce new and technically superior products" (p. 7).

What is technical innovation? To the OECD, "technical innovation is the introduction into a firm, for civilian purposes, of worthwhile new or improved production processes, products or services that have been made possible by the use of scientific or technical knowledge" (p. 9). Technical innovation is also a process in time – with three stages: invention, (initial) innovation ("when a firm introduces a new or improved product into the economy for the first time"), and (innovation by) imitation, later called diffusion (p. 9).

Why did the OECD turn to innovation? Because science policy was only concerned with developing research potential. Yet it is innovation, the whole process, that should be considered the driver of economic progress. Scientific "opportunities will not be exploited effectively unless businessmen have the technical and managerial competence, the incentives and the means necessary in order to innovate" (p. 7). "Technical innovation depends not only on [research and development] but also on the capacity of firms to use its results" (p. 11).

Government and Technical Innovation was not the first of the OECD's thoughts on technical innovation. In 1965, the OECD Secretariat produced an internal document (by an unknown author) summarizing the evidence on factors affecting technical innovation. The document starts by discussing the existing theories on "technical progress" (technical progress as a residual, after capital and labor have been subtracted from the econometric equation) – and the empirical results are reviewed at length. The existing theories (of technological change) are described as being too macro; it remains difficult to break down technical progress into "component parts". In contrast, claims the OECD document, innovation, defined as "the introduction of new and improved processes and products into the economy, or the new application of either new or old knowledge to production processes at the firm or industry level", can be analysed as a "process" or "time sequence" of factors (OECD, 1965: 3, 5).[4] What is important to the OECD is that "this sequence can be shortened [time lag] and the flow of innovation increased by influencing certain factors which create more favourable conditions" (p. 3).[5]

To the OECD, "innovation implies more than invention . . . Innovation is more than the initial putting into production of an object or the introduction of a technique." Innovation is "both a specific process for an individual firm and a flow over time in the broader perspective of the world economy" (p. 5). The technical "potential in itself is not enough, but there must be a recognition or awareness of it" (pp.10–11). "The most efficient means of increasing economic growth [does not] consist uniquely of adding new knowledge to the existing stock . . . [but rather of putting] this knowledge to better use" (p. 12).

After *Government and Technical Innovation*, the OECD continued to produce documents on technological innovation. One such influential study was *Gaps in Technology* – the idea of a "gap" was used at the United Nations in the early 1950s (see Arndt, 1987: 63) concerned with trade issues and market share. In the 1960s, there were concerns in Europe that the continent was lagging behind the United States in terms of technological potential. The OECD conducted a two-year study, collecting many statistics on the scientific and technological activities of European countries and the United

States (OECD, 1968a, 1970a). *Gaps in Technology* contrasted technological innovation with invention (innovation "requires development work, together with manufacturing and market activities"), and defined (measured) technological innovation in terms of Member countries' performance on two aspects: (1) "being first to commercialize new products and production processes successfully" and (2) the "level and rate of [diffusion] of new products and production processes" (OECD, 1968a: 14).

In the end, none of the statistics collected by the OECD appeared conclusive in explaining economic performance. The issue remained part of endless public debate in France and the United States for several years.[6] A huge volume of scholarly literature also developed in the following two decades. The OECD suggested that the causes of the gaps were not related to research and development per se: "scientific and technological capacity is clearly a prerequisite but it is not a sufficient basis for success" (OECD, 1968a: 23; OECD, 1970a). The organization rather identified other factors in the system as causes: capital availability, management, competences, attitudes, entrepreneurship, marketing skills, labor relations, education and culture (OECD, 1968a: 23; OECD, 1970a).

Another OECD document deserving mention, as it was regularly cited at the time, is *The Conditions for Success in Technological Innovation*, written by Keith Pavitt and Salomon Wald. This is the last document Pavitt produced for the OECD. In 1971,

Pavitt joined the Science Policy Research Unit (SPRU) in Brighton, founded by Chris Freeman in 1966. The document was an "immediate sequel" to *Gaps in Technology*. "Technological innovations now create the basis for economic growth" state the authors. "The industrially advanced countries have some sort of collective responsibility to produce useful innovations as a basis for future economic growth" (Pavitt and Wald, 1971: 21–2). To the authors, research and development cannot simply be equated with innovation (p. 24). Innovation is defined as "bringing invention to its first successful commercial use". Again, a triad – invention, innovation and diffusion – is suggested (p. 19):

> Invention is the idea of how science and technology could be applied in a new way, innovation consists of bringing invention to its first successful commercial use, and diffusion consists of the spread of the use of the innovation amongst its potential population of users.

Although "an artificial oversimplification", the distinction is "indispensable . . . to understand the *various* [my italics] economic policy implications of science and technology" (p. 19). The study suggests a "model" of innovation as a process composed of three factors, as Sumner Myers put it: knowledge, demand (or market needs) and the coupling (of the two), presented in an "institutional" and "system" framework: the *"total* [my italics] innovative system" is composed of three parts or institutions acting together, namely industry, universities and government (p. 22). "National fundamental research and national innovative capabilities form part of the same national system" (p. 104). The study is wholly concerned with surveying the role and the conditions or factors of each institutional sector in the innovation system.

The US Department of Commerce

An influential input into these views came from the US Department of Commerce. In 1964, the US President asked the Department of Commerce to explore new ways of "speeding the development and spread of new technology". To this end, Herbert Hollomon, Secretary for Science and Technology, set up a panel on invention and innovation, whose Chairman was Robert Charpie (President at Union Carbide Electronics) and whose Executive Secretary was Daniel de Simone, an electrical engineer and Director of the Office of Invention and Innovation at the National Bureau of Standards. The report, known as the Charpie report, was published in 1967 as *Technological Innovation: Its Environment and Management*. From that time on, technological innovation replaced technical invention as a key term in public documents and elsewhere.

"Why should the government have an interest in invention and innovation?" asked the report. "The answer is that invention and innovation lie at the heart of the process by which America has grown and renewed itself" (US Department of Commerce, 1967: 2–3). "To be sure, innovation is not limited to technological products and processes in the business world. But that is the principal sense in which we were asked to be concerned with innovation" (p. 2). Technological innovation is the source of economic growth, international leadership (trade) and competitiveness (pp. 3–7).

The report begins by making a distinction between inventing and innovating: the difference being between the verbs "to conceive" and "to use".[7] Inventing is to conceive an idea, and innovating is to use the idea, or the process of translating the idea into the economy. "We find that the concepts, uncertainties, and other realities of technological innovation are like a foreign language, indeed a strange world, to too many of us. Because of this, we believe the most important initial task before us is to become more widely acquainted with the 'language' and 'world' of innovation" (pp. 56–7). To the Department of Commerce, innovation means use, not conception or generation: innovation is a "complex process by which an invention is brought to commercial reality" (p. 8). Again, research and development is only one stage or phase or step (the vocabulary is rather broad) of this process. Innovation includes research and development, engineering, tooling, manufacturing and marketing. Using "rule of thumb" figures from the "personal experience and knowledge" of members of the panel, the Department of Commerce reported that research and development corresponds to only between 5 per cent and 10 per cent of innovation costs. "It is obvious that research and development is by no means synonymous with innovation" (p. 9).

These numbers were regularly cited in the years that followed, not only in the United States but also among European officials.[8] Hollomon, Charpie, Simone and their consultants played the role of ambassadors, disseminating the message to many audiences (Hollomon, 1965a, 1965b, 1967, 1968; Charpie, 1967a, 1967b, 1970; Simone, 1965, 1967, 1974; Michaelis, 1964, 1967; Gellman, 1966, 1970). The Department of Commerce paved the way for an influential representation of innovation in the following decades. Policymakers, supported by engineers, managers and scholars, embraced this representation without reservation. Technological "innovation is not simply research and development" (US Department of Commerce, 1967: 8). It is a "total process", an "entire venture", embedded in a "total environment" (p. 2; see also pp. 8, 11, 14):

> Invention and innovation encompass the *totality* [my italics] of processes by which new ideas are conceived, nurtured, developed and finally introduced into the economy as new products and processes; or into an organization to change its internal and external relationships; or into a society to provide for its social needs and to adapt itself to the world or the world to itself.

To the Department of Commerce, innovation was a "total process" of several activities in *firms*, together with appropriate government legislation, or a "complex system" to address competition issues. The factors of interest to the Department of Commerce were the structure of industry and competition (firm size) and "barriers" to innovation (government regulations). Therefore, the recommendations were concerned with tax legislation (on research and development), venture capital and antitrust law, as a report on technical innovation from the US Chamber of Commerce had suggested the year before.[9] Research, per se, had no place in the committee's reflections.

Again and again, the committee responsible for the report stressed that we knew too little about the process of technological innovation – a motto found in every official document of the time (pp. 30, 45, 56–7). There was an "abundance of ignorance about the processes of invention, innovation and entrepreneurship". As a consequence, the committee recommended the initiation of a research program on innovation and entrepreneurship, a high-level conference, and a series of regional conferences "aimed at removing barriers to the development of new technological enterprises, jobs, and community prosperity in the respective regions".

The Department of Commerce report, "a nonimplemented 'classic' in this field", claimed a survey from the Research Congress Service in 1980 (US Joint Economic Committee, 1980: 12) was not the first US government thinking on innovation. As discussed previously, in 1962 the National Science Foundation (NSF) commissioned a report on "technical innovation" from consulting firm Arthur D.

Little Inc., a report with the same representation of innovation as that which the Department of Commerce and the OECD espoused later: innovation is a "process of bringing invention into commercial use" (Arthur D. Little Inc., 1963: 6). The main message of the report is that technology is not enough. It is the use of technology that counts. In 1965, the Senate held hearings on *Concentration, Invention and Innovation*, at which many economists such as Richard Nelson, Frederic Scherer and Jacob Schmookler testified, as did Simone. Senator Philip Hart opened the hearings, stating "It is important to have a clear conception of what we are talking about." The problem involves "a confusion of the concepts of invention, on the one hand, and innovation or development on the other hand". Innovation is "a prolonged and very expensive process [namely] the development of an idea into practical form for efficient production and distribution" (US Senate, 1965: 1074). Senator Hart's comments had little success. The invited speakers talked of invention (and research) rather than innovation.[10] The following year, the NSF, together with the National Planning Association, from which it commissioned what became an influential study of the innovation process (Myers and Marquis, 1969), organized a conference on technology transfer and innovation. Hollomon, Charpie and Simone participated as speakers (US National Science Foundation, 1966). That same year, the Department of Commerce produced an annotated bibliography on the subject, which espoused, again, the representation of innovation as the commercialization of an invention: "Innovation may be described as the introduction of new techniques into a company, an industry, a market, or a region" (US Department of Commerce, 1966: 13). The document attributed this definition to Hollomon – as another scholar did later (Goldsmith, 1970: xiv).

The British Central Advisory Council for Science and Technology

One year after the US Department of Commerce report, the British Central Advisory Council for Science and Technology (CACST), an organization established the year before "to advise the Government on the most effective national strategy for the use and development of our scientific and technological resources", produced *Technological Innovation in Britain*, known as the Zuckerman Report (from the name of the Council Chairman, scientific adviser Solly Zuckerman, a pioneer of operations research). As soon as it was published, the report was debated at length. It recommended, among other things, reducing public support for research that had no "foreseeable [industrial] production", diverting resources to innovation activities, and "running down" public establishments that "cannot provide service for industry" (British Central Advisory Council for Science and Technology, 1968: 16–17). None of the reports studied above had gone that far.[11]

Like the OECD and the US Department of Commerce, the CACST's rationale for looking at innovation was economic: "A major purpose of technological innovation is the commercial exploitation of technical knowledge so as to win new markets or to hold existing ones against competition, and to reduce costs of production" (p. 1). Like the OECD and the US Department of Commerce, the report's main purpose was considering the factors that determine technological innovation, which add up to five: research and development, planning, management, lead time and market structure (scale of production and size of market).

Like that of the US Department of Commerce, the report starts with a discussion of what technological innovation is (p. 1):

> The term "technological innovation" can be defined in several ways ... At one extreme innovation can imply simple investments in new manufacturing equipment or any technical measures to improve methods of production; at the other it might mean the *whole* [my italics] sequence of scientific research, market research, invention, development, design, tooling, first production and marketing of a new product.

The CACST report defined technological innovation as a process: "the technical, industrial and commercial steps which lead to the marketing of new manufactured products and the commercial use of new technical processes and equipment". As with the OECD and the US Department of Commerce, to the CACST, innovation is a sequence or "chain" of related steps, as one member of the CACST put it in *Nature*: "pure science, applied science, invention, development, prototype construction, production, marketing, sales and profit (Blackett, 1968: 1108). "R&D, production and marketing . . . constitute a *single* [my italics] innovative activity" (British Central Advisory Council for Science and Technology, 1968: 3). "A high level of R&D is far from being the main key to successful innovation" (p. 9). "The most difficult and complex problems in the process of technological innovation generally lie in this final phase [of marketable products which the customer wants and the producer can make at a profit], the phase which includes aggressive and sophisticated marketing" (p. 15). The report concluded that Britain needed to achieve a "new balance": "Government support should be given to the *whole* [my italics] process of technological innovation, in contrast to its present overwhelming emphasis on the opening phases of research and development" (p. 15).

From the beginning, the report placed the emphasis on the leading position of the US and the undervaluing of innovative activities in Britain. This is an old issue, as mentioned above.[12] "One major strength of American industry has lain in its ability to carry an idea through to the final product without a break in the innovative chain . . . The most successful American corporations mostly dominate in the final stages of innovation" (pp. 3, 6). In contrast, Britain espouses an "old practice . . .

of treating R&D as something apart from the rest of the industrial process" (p. 11). "The evidence is clear that on a national scale the attractions to the professional scientist and engineer of engaging in research and development have detracted greatly from those of other spheres of industrial activity. A bigger fraction of qualified manpower must be encouraged to go into the management of industry than at present" (p. 14). The advantage is "in the final stages of the process of innovation, in tooling up, in introducing improvements in processes, in skillful and aggressive marketing" (p. 5).

In the same year as *Technological Innovation in Britain*, the CACST commissioned another study to explore "why Britain appears to be obtaining a poor return from its large investment in qualified scientists and engineers". The study compared British and American companies and competitors on the European continent in ten industries (see Layton, 1972, for a published version of the report). The message, again, was that innovation (production, marketing and entrepreneurship), while "not accepted as a top priority on a sufficient scale" in Britain, is the key to success.[13] The study offered ways to improve the situation through "change in basic national social values" (attitudes), education and management.

The National Research Development Corporation

Members of the British National Research Development Corporation (NRDC), created in 1948, were active participants in this debate. John Clifford Duckworth, an engineer and Managing Director of the NRDC from 1959 to 1970, is one example. Duckworth's discourse preceded the CACST report by a few years. In a Graham Clark Lecture on 24 March 1965 before the Institute of Mechanical Engineers (*Incentives to Innovation and Invention*), Duckworth discussed the British attitude to the exploitation of research results: "It is often said that as a nation we are good at fundamental research and invention but not so good at putting results to profitable use; for many years there has been too great a tendency for the better men to turn to pure science" (Duckworth, 1965a: 190). At the time, British discoveries (and others from European countries) were being exploited commercially and patented in the US, so it was claimed (British Advisory Council on Scientific Policy, 1964: 8; British Central Advisory Council for Science and Technology, 1968: 1, 3, 6). This "provided a major argument in favour of setting up the National Research Development Corporation" (Duckworth, 1965a: 187).

Duckworth claimed that "the future national welfare of our country depend[s] largely on the speed with which industry could turn to new, commercially viable, processes and products . . .

[Yet] the pure scientist appears to be held in higher esteem than the engineer and technologist ... [There is a] lack of status of the professional engineer as compared with the scientist" (p. 186). To Duckworth, "inventions and innovations are not necessarily meritorious in themselves, but only in so far as they contribute to higher efficiency and enable us to compete more effectively in world markets" (p. 188). Innovation is "the application of the invention or new technique" in industry (p. 187), and this is the task of the manager. "I have no regrets whatever at having deserted the more academic scientific pursuits, and I would advise any young scientist or engineer, who has other than purely academic abilities to move unhesitatingly towards application and management. In my view, it is wrong to say – as is often done – that it is a waste of a scientist when he enters management" (p. 186). Duckworth urged a "reversal of this trend", a "change of outlook", a "reorientation of our sense of values" (p. 190). In brief: the "creation of a society which welcomes innovation" (ibid.). Prophesying a bit, Duckworth added, "Perhaps one of the most helpful contributions we could ask from the classicists is that they should coin a new and socially acceptable single word to replace the clumsy expression 'Chartered Engineer'" (p. 186). "Innovator" did the job. History demonstrates that the concept of innovation offers an organizing and mobilizing idea for what Duckworth calls a "reorientation of our sense of values".[14]

In another lecture of the same year, before the Electronic and Radio Engineers in Southampton, titled *The Process of Technological Innovation* (25 February 1965), Duckworth suggested four "requirements for innovation": a clear national objective, the coupling of scientific possibilities with the objective, the full exploitation or commercialization of fundamental science, and public support to industrial financial and manpower resources (Duckworth, 1965b).

At a conference organized by the Science for Science Foundation in April 1969, Duckworth discussed the role of government in "encouraging technological innovation" and economic growth. He identified two kinds of measures, called "direct" (defense; purchasing policy and funding applied research; education; development of civil aircraft and atomic energy; retraining programs; exploiting pubic inventions) and "indirect" (patent system, taxation, spin-off). In his talk, Duckworth continued his previous discourse on the "innovating society". "In this country there is little doubt that in the post-war years too much emphasis was placed on 'science' and more latterly 'technology', without a full appreciation of the financial and other resource needed for engineering design, production and marketing of products and processes . . . Our real need is for a large proportion of our best people to have a general training, including science and engineering and then to become business entrepreneurs" (pp. 114–15). "It is industry which in general has the knowledge of markets, and it is industry which should best be suited to select future applied research and development areas. Government must first create the environment" (p. 116).

Notes

1. Words used include: innovation, technical innovation and product/process innovation.
2. "A rise in output may be expected to depend increasingly on the rise in productivity resulting from scientific research and technological development" (p.19); "Growth industries are mainly those in which major technical innovations take place as a result of new scientific discoveries" (p.29); "The financial burden of research is often the most important obstacle to innovation" (p.55); "Most innovations are based on scientific discoveries" (p.63).
3. Education, public research, procurement – government as "enlightened buyer" – support to industry – "which are of basic importance to future economic development" – taxation, patent system, and so on.
4. The OECD document rejects what the author calls "the traditional approach of Schumpeter" (the sequence invention, innovation, imitation) because it is limited to the "first new" and does not consider later improvements, and the document offers a sequence from initial idea to commercial production, with "interdependent steps": research, development, diffusion, improvements.
5. Five factors affecting innovation are studied: research and development, knowledge from outside (patents and licenses), talent, market structure and investment (capital).
6. The debate was initially launched in France in the early 1960s, with statistics on the technological balance of payments that showed that the United States was receiving ten times as much as what it was paying to

other OECD Member countries (see Godin, 2005: ch. 12).

7. The phrasing most probably comes from Thomas Whisler, Professor of Industrial Relations, Graduate School of Business, University of Chicago (see Becker, 1967: 2).

8. Both the OECD report on technological gaps (OECD, 1970: 188) and the British Central Advisory Council for Science and Technology report, discussed below (British CACST, 1968: 7) cited the report.

9. "Laws, regulations and practices ... too often hinder private investments which are needed in our expanding technological economy ... The atmosphere should be as free as possible from restraints or barriers to innovation" (US Chamber of Commerce, 1966: Foreword, 1).

10. Speakers assumed development and patents to be synonymous with innovation, and did not consider introduction (use) and commercialization, with the exception of Daniel Hamberg and Daniel de Simone.

11. A few years before, an American scholar proposed a similar scheme for increasing private research and development: "An ingenious method for inducing an increase in private expenditure on research and development with no additional cost to the federal government would be to reduce federal expenditures on research and development each year by a stated amount and to offer this amount to industry" (Warner, 1966: 216).

12. In the United States, the view was the opposite. That the United States is best in applied research but lacks basic research is an old thesis, going back to the nineteenth century. On a critique of this interpretation of research in America, see Reingold (1991). The view also served as an argument for launching science policy in Europe in the early 1960s (see Godin, 2002).

13. "Innovation ... means not merely inventing, but also developing and producing the results of inventions and bringing them to the market place" (p.xi).

14. Another member of the NRDC who spoke in similar terms was Brian Locke, engineer and head of special projects, in a lecture delivered at Imperial College in 1976. "Invention, which is discovering ways of doing things unexpectedly . . . is only one of many components of innovation, and of negligible use without design . . . Innovation is not r. & d. although it begins with research, and continues with . . . development" (Locke, 1976: 611–12).

5. Innovation as system

By the late 1960s, a discourse on technological innovation was developing, and practitioners were key participants. The discourse gave rise to a new semantic pair or dichotomy: research–innovation. Research is not a prerequisite to technological innovation. That view evolved in only a few years. For example, at a conference in 1962 Herbert Hollomon claimed that "science is the source from which new technology derives, and science is crucial to [innovation]" (Hollomon, 1965a: 254). In 1967, he had changed his view: "most technological change, most innovation, most invention, and most diffusion of technology are stimulated by demand . . . and only indirectly science-related" (Hollomon, 1967: 34). Similarly, Keith Pavitt argued in 1963 that "fundamental research is, in the long run, an essential pre-requisite for innovation and economic growth" (Pavitt, 1963: 209). The document he wrote for the OECD ministerial conference in 1966 suggested otherwise: "technical innovation depends not only on R&D, but also on the capacity of firms to use its results" (OECD, 1966: 11). To Pavitt's colleague Chris Freeman, the "crucial step" is commercialization (Freeman, 1963), a view opposite from the background document he produced for the OECD in 1963 (OECD, 1963b: 63).

The new discourse rests on four criticisms of the then-emerging view of technological innovation. The first concerns the linearity of the process. Early emblematic critics include William Price and Lawrence Bass, formerly of the US Department of Defense and Arthur D. Little Inc. respectively, to whom "The 'linear model' is not typical, as Donald Schon from Arthur D. Little Inc. put it a few years before. One appreciates the non-rational nature of the innovative process when one notes that the more novel the invention is, the less orderly . . . is the process" (Price and Bass, 1969: 803). Yet every writer, with few exceptions, admits qualifications. The policymakers studied in Chapter 4 never believed that technological innovation is a strictly linear process, from basic research to commercialization, and said so explicitly in every policy document. As engineer Jack Morton from Bell Laboratories put it, "It is useful to talk of the innovation process as if it were an orderly sequence, always remembering that the ordering and timing of the various parts are neither rigid nor done only once" (Morton, 1971: 19–20).

A second criticism suggests that the initiating stage or fundamental factor in technological innovation is not science or scientific research. Already in the 1930s, diverse studies said the same of invention. American patent examiner Joseph Rossman argued in *Psychology of the Inventor* that "it has been frequently repeated that science is solely responsible for the technical advances in our industries. This statement has been made so often that most people accept it as axiomatic. Actual facts, however, do not warrant the assumption" (Rossman, 1931: 18–19). To sociologist Colum Gilfillan, science is "non-necessary" for invention. Gilfillan's fourth principle of *The Sociology of Invention* claims that "invention need not be based on prior science" (Gilfillan, 1935: 6).

The discourse of the 1960s and 1970s is similar but less extreme. A 1968 symposium sponsored by the US National Academy of Engineering concluded "there appears to be general agreement that the process of successful technological innovation depends on many more factors than the mere generation of scientific and engineering information" (US National Academy of Engineering, 1968: Foreword). Managers thought similarly. The summary statement of the annual meeting of the Industrial Research Institute (IRI) on innovation, where more than 100 industrial research managers gathered in April 1970, begins with the following "authoritative picture" of innovation: "Innovation is the process of carrying an idea – perhaps an old, well known idea – through the laboratory, development, production and then on to successful marketing of a product . . . [Yet] the technical contribution does not have a dominant position" (Research Management, 1970: 45).

Numbers seconded the argument. Research amounts to a small percentage of innovation activities, so measurements showed. The statistics came from both practitioners (US Department of Commerce, 1967) and scholars (Mansfield et al., 1971; Mansfield and Rapoport, 1975; Mueller, 1957).[1] In a literature review produced for the OECD in 1971, Keith Pavitt concluded: research and development statistics "measure only one part of the total input into the innovative process . . . research and development expenditures cannot be equated with innovative expenditures" (Pavitt and Wald, 1971: 24).[2]

A third criticism claimed that demand or needs are more fundamental than research (Godin and Lane, 2013). As the Department of Defense had, practitioners began to look at technological innovation from a demand rather than a supply perspective, arguing that the most critical element in technological innovation is need-pull forces (opportunities *pulling* from peoples' needs and the market) rather than supply-push forces (technological opportunities *pushing* forward from scientific discoveries). To be sure, many scientists and scholars continued to contest the role of needs or market demand in technological innovation, often in polemical terms (see Kreilkamp, 1971; Mowery and Rosenberg, 1979). Yet needs contributed to the inclusive view of technological innovation that we hold today. Herbert Hollomon's discourse is a perfect example of the rhetoric of the time. "Most technological change, most innovation, most invention, and most diffusion of technology are stimulated by demand . . . and [are] only indirectly science-created" (Hollomon, 1967: 34).[3]

Hollomon summarized his idea in his speech to a conference on the economics of research and development held at Ohio State University in 1962 (Hollomon, 1965a: 253):

> The sequence – new science from research, application of new science, development, prototype manufacturing, and sales – is not the usual way innovation occurs. The majority of new processes that increase our ability to turn out products and services efficiently, broaden our economic life and widen our variety of choice, take place as a result of a process that involves the recognition of a *need*, by people who are knowledgeable about science and technology. The sequence – perceived *need*, invention, innovation (limited by political, social, or economic forces), and diffusion or adoption (determined by the organizational character and incentives of industry) – is the one most often met in the regular civilian economy.

Fourth, "research is not enough" – a key phrase of the 1960s – in the sense that it requires use or adoption.[4] "Invention without innovation may be intellectually satisfying, but it does nothing to promote the general welfare", claimed Daniel de Simone. "As far as society is concerned, invention without innovation is an unwritten poem, an unpainted picture, an unplayed symphony" (Simone, 1965: 1095–6). Equally, Aaron Gellman, Vice-President of the North American Car Co., and member of the panel responsible for the US Department of Commerce report of 1967, wrote (Gellman, 1970: 136):

> It is well to note also that there are a number of popular notions about innovation which do not stand up to

the cold light of an analysis of the process of innovation. Specifically, the concept that if one builds a better
mousetrap the world will beat a path to one's door is a
dangerous one upon which to base decisions relative to
the allocation of scarce resources to innovative activities
. . . Similarly, the notion that "nothing can stop an idea
whose time has come" is absurd.

To Gellman "in most cases market analysis and
marketing take a significant portion" of the innovation costs. But this is too little appreciated by
innovators, policymakers and scholars (1970: 129).

This message has old origins. Discovery or
invention "is without result and sterile unless it is
adopted", stated anthropologist Roland Dixon in
1928. "Without its diffusion beyond the discoverer
or inventor, the new trait remains merely a personal eccentricity, interesting or amusing perhaps,
but not significant" (Dixon, 1928: 59). Innovation
is use or adoption, that is, action rather than con-
templation or speculation. The policy documents
studied in Chapter 4 reflect the same view.[5] To
most practitioners, use or application means by
the market. Such too is the understanding of
scholars.[6]

All in all, innovation is a "total process", from
conception to application. Together with Donald
Schon, who observes a "shift away from an uncrit-
ical acceptance of research as a means to growth"
(Schon, 1967: 56), one of the most ardent publicists
of the holistic nature of technological innovation
at the time was Jack Morton. Innovation is "the
totality [my italics] of human acts by which new

ideas are conceived, developed, and introduced"
(Morton, 1971: 2–3):

> Innovation is not just one single act. It is not just a new
> understanding or the discovery of a new phenomenon,
> not just a flash of creative invention, not just the develop-
> ment of a new product or manufacturing process; nor is it
> simply the creation of new capital and consumer markets.
> Rather, innovation involves related creative activity in *all*
> these areas. It is a *connected* process in which many and
> sufficient creative acts, from research through service,
> couple together in an integrated way for a common goal
> ... By themselves research and development are not
> enough to yield new social benefits. They, along with
> capital resources, must be effectively coupled to manu-
> facturing, marketing, sales, and service. When we couple
> all these activities together, we have the connected spe-
> cialized elements of a *total* [my italics] innovation process.

The four criticisms discussed above crystallized
into the concept of system. Certainly, process
signifies a system or "total" set of activities, in
time. But system adds one more dimension: space.
System was a popular concept in the 1950s–60s
(Talcott Parsons, *The Social System*, 1951; David
Easton, *The Political System*, 1953; Karl Deutsch,
The Nerves of Government, 1963) – and *system
dynamics* was another (Jay Forrester, *Industrial
Dynamics*, 1961; Ludwig von Bertalanffy, *General
System Theory*, 1968) – first used in ecology in the
1920s and in cybernetics in the interwar years.
The concept entered into studies of science and
technology in the 1960s, particularly in the United
States at RAND (Hounshell, 2000). Many scholars,
particularly from management schools, began

to use a system approach to study decisions and choices regarding science, technology and innovation. The view that the firm is a system was shared by many at the time (see Burns, 1975; Burns and Stalker, 1961; Morton, 1971; Mueller, 1971). The concept shares a place with the term 'matrix' in the vocabulary of the time (e.g. Caffrey, 1965; Hertz, 1965). The US Department of Commerce and the OECD began to apply the idea at the national and policy levels.

The US Department of Commerce

In 1968, the Office of Invention and Innovation of the US Department of Commerce, at the instigation of its Director Daniel de Simone, and as a follow-up to *Technological Innovation: Its Environment and Management* (US Department of Commerce, 1967) commissioned a study from Ellis Mottur, senior staff scientist in the Program of Policy Studies in Science and Technology at George Washington University.[7] The result of the study is a report titled *The Processes of Technological Innovation: Conceptual Systems Model* (1968).

To Mottur, technological change requires a "conceptual framework", an "analytical tool" that can assist in predicting, directing and structuring the phenomenon. To this end, Mottur produced a system model that "attempts to take account of the full range of factors and interactions involved in the processes of technological innovation" (Mottur, 1968: 29). According to Mottur (1968:

189), the chief significance of the model is that it is:

> A single systems model of the processes of technological innovation which spans the full spectrum of innovative processes from basic research through to use and appraisal in the socio-economic system; which is applicable to all kinds of innovations (specific products, composite products, processes, services, and systems); and which permits consideration of all kinds of institutions, organizations, groups, and individuals that may be involved in innovative processes . . . It provides a coherent framework of interrelated functions, within which one can systematically describe and analyze the many variations in the processes of technological innovation as they occur in practice.

Mottur admits that "the word system is in reality something of a misnomer". There is no system in the sense of "constituting a coherent *whole* [my italics] that functions in some sort of interrelated manner" but an "exceedingly complex maze of institutions that interact – when they do interact – in oftentimes ineffectual or confusing ways" (Mottur, 1968: 49–50). Mottur's model is "a flow system in which goals, resources, technological knowledge, and related information flow through varying networks of institutions, organizations, groups, and individuals . . . to produce the technological innovations that enter the socio-economic system" (pp.168–9).

The model is composed of 61 steps or "functions" plus more than 200 sub-functions and problem areas (for the Figure, see Godin, 2019).[8] It maps technological activities and their relationships to the socio-economic system, as well as to the educational system and what he calls the conversion process. Conversion is a key term to Mottur. There is an "internal technological gap" between technological innovation and the socioeconomic system, an idea not dissimilar to that of sociologist William Ogburn's cultural lag (Ogburn, 1922). Mottur calls the gap "internal, to distinguish it from the [then-highly-popular concept of the] technological gap between countries resulting from their differing technological possibilities" (Mottur, 1968: 65). Reducing the gap requires conversion activities, namely adapting technological activities to societal needs.

In the end, Mottur's system model did not circulate much. The report remained part of the gray literature, garnering only a couple of citations. Yet the report is witness to the then-increasing appropriation of a system approach to innovation. For example, to Donald Schon, "there is no single institution or set of institutions. There is only the whole complex of institutions: companies, industry associations at varying levels, universities, research institutes and governments" (Schon, 1967: 173). Schon developed the idea of a "learning system", namely social systems that "learn to become capable of transforming themselves without intolerable disruption" (Schon, 1971: 60). Jack Morton elaborated a "system model of innovation" too, made up of interrelated parts, namely "people . . . coupled together" (Morton, 1971: 15–16). Using the "systems [sic] approach" of engineers,[9] biological ecology and the Bell system as model, Morton defined a model as: "a system is an integrated assembly of specialized parts acting together for a common purpose ... a group of entities, each having a specialized, essential function. Each is dependent for its system effectiveness upon its coupling to the system's other parts and the external world" (Morton, 1971: 12–13).

Both Schon and Morton were members of a network parallel to that of the US Department of Commerce (Herbert Hollomon, Robert Charpie, Daniel de Simone, Aaron Gellman), the short-lived Innovation Group (1969–72), whose objective was to promote innovation and understanding of innovation to diverse publics (Wisnioski, 2012: 148–58). One instrument was *International Science and Technology*, a monthly journal founded in 1962, which changed its name to *Innovation* in 1969 – certainly the first journal ever with the concept of innovation in its name.[10]

The OECD

Starting at its creation in 1961, the OECD produced several policy documents, and most of them carried a system approach (Godin, 2017). The organization studied research and scientific and technological activities ("science and technology", as it was called) as a system, with the term as such,[11] a system itself embedded in a larger system or environment (OECD, 1960, 1963a, 1963b, 1966, 1968a, 1974b, 1980). Pavitt and Wald's study of 1971 applied the idea to innovation (see Chapter 4).

Then in 1972, the OECD Directorate for Scientific Affairs created an ad hoc group on industrial innovation. Between 1972 and 1978, the group collected information on policy instruments or measures of technological innovation in Member countries. The survey was conducted using two conceptual frameworks that served to classify the measures. These frameworks were conceived of as early as 1972. One was according to *time* (process), as the organization called it, or stages of the innovation process (the linear model of innovation), and the other according to *space* (system), or the institutions involved in the process. For each framework a diagram was included, that on institutions being similar to those used today to visualize national innovation systems (OECD, 1972a: 12; OECD, 1978: 142) (see Figure 5.1).

The system approach was adopted for policy concerns early on (see Part II of this volume). It gives social life or existence to policy measures of an "integrating" kind, in comparison to the measures directly connected to the process of innovation (stages) *in* enterprises. Integrated policy measures are concerned with the *relationships* between institutions to facilitate the transition from the invention stage to the marketing stage: the transfer of knowledge from public to private organizations, spin-offs from public research programs to industry.

Early on, an OECD report identified some

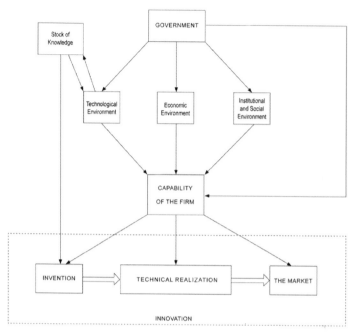

FACTORS INFLUENCING INNOVATIONS

Figure 5.1:   *Factors influencing innovations*

problems with the notion of system. "It is not enough to say that everything depends on every-thing else, and so we must look at the whole system". This is a "counsel of despair", a "pretext for inaction" (OECD, 1971: 57). Be that as it may, in the following decades, scholars embraced the system approach to innovation and made of it a key characteristic of "innovation studies": "The notion of national systems of innovation is one of the most important developments to emerge from innovation studies in the last 25 years", claim some scholars (Fagerberg et al., 2013: 173).

However, this should not be interpreted to mean that the process approach has disappeared, quite the contrary.[12] Both the process and the system approaches equally continue to serve policy – and scholarship.

According to scholars, the concept of "system of innovation" comes from academics in the late 1980s. However, a careful study of history shows that the concept goes back to the 1960s, including the first uses of the concept at the national level. History also reveals a voluminous study of 1980 from the US Congressional Research Service (CRS) submitted to the Joint Economic Committee of Congress that analyses innovation as a "holistic" process. The CRS study represents the first public use of the expression and examination of "innovation system" at the national level. "We have begun to see" claimed the analysts participating in a workshop in 1978 that launched the CRS study "the systemic and holistic nature of [innovation] processes" (Mogee and Schacht, 1980: 13–14). The study suggests that there is a "necessity to view innovation in the system/subsystem contexts", namely at "three integrative levels": socioeconomic, functional (elements such as capital, competition, research, intellectual property) and industrial (US Joint Economic Committee, 1980: 2, 4). In spite of a "hubbub of innovation writings and discussions", claimed the committee, there are no comprehensive findings and no integrative policies (p.6).

Notes

1. Mansfield's numbers showed "substantial agreement" with the US Department of Commerce, although research and development expenditures were higher to Mansfield (17.1 per cent versus 5–10 per cent). Mueller's number is 7.4 per cent.
2. The statistics did not remain uncontested (see Stead, 1976).
3. For scholars' views, see Langrish et al. (1972), Myers and Marquis (1969), Rothwell and Robertson (1973), SPRU (1972), Utterback (1974).
4. The study of this discourse is generally not part of the scholarly literature on the history of science and technology, nor that of science and technology policy. This may explain why some historians do not see the linear model of innovation in the literature (for example, Edgerton, 2004) or why most others attribute the origins of the model to heterogeneous sources. In fact, the linear model is the cumulative results of many theorists, both practitioners and scholars, over many years, even decades (Godin, 2017). One has to turn to the literature on technological innovation to document the discourse. The study of science and technology and the study of technological innovation are two distinct fields – although science (basic research) is postulated as being at the origin of technological innovation.
5. "The factors involved are by no means all, or mainly scientific; some of the most important are indeed socio-logical" (British Advisory Council on Scientific Policy, 1964: 8); "a high level of R&D is far from being the main key to successful innovation ... Government support should be given to the whole process of technological innovation, in contrast to its present overwhelming emphasis on the opening phases of research and development ... The most difficult and complex problems in the process of technological innovation generally lie in this final phase [of marketable products that the customer wants and the producer can make

at a profit], the phase which includes aggressive and sophisticated marketing" (British Central Advisory Council for Science and Technology, 1968: 9, 15). "Technical innovation depends not only on R&D, but also on the capacity of firms to use its results" (OECD, 1966: 11). "Scientific and technological capacity is clearly a prerequisite but it is not a sufficient basis for success" (OECD, 1968a: 23). "It is not enough for the inventor to invent; he must also bring his idea for a new product or process to market" (US Advisory Committee on Industrial Innovation, 1979: 6). "The most intractable problems lie not in the potential of science and technology as such, but rather in the capacity of our economic systems to make satisfactory use of this potential" (OECD, 1980: 93).

6. There is a difference between "scientific gestation" ("time elapsing between the first conception of an idea and its public presentation") and "industrial gestation": "the time elapsing between the end of the scientific gestation and the date when an innovation has in an economic or industrial sense 'arrived' or been fully accepted", namely the "date when a thing is available on the market" and particularly "when it has economically justified itself" (Stamp, 1929: 93). "Economic progress is the orderly assimilation of innovation into the general standard of life. I use the word 'innovation' advisedly, for 'invention' has too mechanical a connotation, and we must include processes and the consequences of discovery" (Stamp, 1933: 383). "It is only when they [inventions] can be commercially used that they are of any importance to market considerations. Until that time, they may exist technically, but they do not exist economically" (Lederer, 1938: 32). "Innovation is possible without anything we should identify as invention and invention does not necessarily induce innovation": invention is an act of intellectual creativity and "is without importance to economic analysis", while innovation is an economic decision: a firm applying or adopting an invention (Schumpeter,

1939: 84–5). "Innovation requires not only the inventor who conceives a product or service to meet a customer's needs, but also the entrepreneur who underwrites and markets the new product" (Warner, 1965: 211). "A great deal of technology" does "not depend on science ... [but] draws on it incidentally ... Most inventions were not made by scientists ... Invention is still the basic ingredient of innovation ... Having a new idea and demonstrating its feasibility is the easiest part of introducing a new product. Designing a satisfactory product, getting it into production, and building a market for it are much more difficult problems" (Wiesner, 1966: 11–12, 15).

7. Mottur graduated in Business Administration in 1954 (master's degree from Harvard Business School). He worked at the National Science Foundation in the 1960s, then became Assistant Director of the Office of Technology Assessment in the 1970s, and thereafter worked as Deputy Assistant Secretary for Technology and Aerospace at the Department of Commerce until 2001.

8. Why so many steps or functions? Mottur is concerned with what he calls the "processes" of innovation, in the plural. "The singular term process ascribes too much simplicity and uniformity to a complex course of events" (p. 11). Innovation is a "complex sequence of interactions" (p. 16), "a set of interrelated processes, rather than a collection of disconnected activities" (p. 12).

9. An analogy made in the late 1930s by Edna Lonigan, US Department of Labor, before she joined Brooklyn College (New York) as Professor of Economics in 1942: "Engineers would not think of making a radical change in design of part of a power system without even looking to see what changes might be necessary in the rest of the system. Architects would not redesign one wing of a house and not expect to consider at all the lines of the rest of the house. In engineering and architecture the organization of separate ideas and

judgments into a functioning whole is recognized as the ultimate responsibility. That stage has not yet been reached in political economy" (Lonigan, 1939: 256).

10. In 1972, the journal became *Business and Society*.
11. Jean-Jacques Salomon's "international scientific policy system" is perhaps one of the first uses of the concept at the OECD (Salomon, 1964).
12. One could argue that it is rather the system approach that has reached its limits. For example, the OECD recently abandoned the approach as a framework for the measurement of innovation (OECD, 2018b).

PART II

FROM IDEA TO ACTION

Practitioners, having defined technological innovation, turned to constructing policy. Construction
involves two acts. One, imagining what an innovation policy is. Two, developing policies. To
some extent, innovation policy existed before the
discourse on technological innovation. Innovation,
defined as a process, covers stages from basic
research to application and commercialization;
policy measures for each of these stages have
existed for some time: for research, patents, education, trade, taxation, and so on. The most obvious
case is research policy. Innovation is defined as
a process that includes research as a first step
or stage. Research policy measures began to be
implemented after the Second World War, and
even earlier in some cases. Equally, the idea of the
commercialization of research results preceded
innovation policy as such. However, innovation as
the commercialization of research, and the invention of research and innovation policy in this sense,
appeared much later, namely in the 1980s.

What was missing until then was a name for
this series of measures, a framework and a coordination of the actions. Our task in Part II of this

book is to understand when innovation policy came to be. What distinguishes science (research) policy, science and technology policy, and innovation policy? When did innovation policy begin to include activities from research to commercialization in the same set of measures, thus blurring the centuries-old distinction between an idea and its application, between research and use, between invention and innovation, leading a majority of people in the following decades to continue talking of innovation in terms of research, in spite of a parallel discourse on technological innovation as commercialization?

Until the 1970s, the measures of what came to be called innovation policy constitute what was called Science, then Science and Technology (ST). Over the years, ST became STI (Science, Technology and Innovation). STI integrates science policy, technology policy, economic policy and others policy measures, to the extent that the measures influence technological innovation. Emblematic of the change in the vocabulary is the OECD Directorate for Scientific Affairs (1961), which changed its name to the Directorate for Science, Technology and Industry in 1974 (I standing for Industry, rather than Innovation as many acronyms had it). The OECD Committee for Scientific Policy also changed its name, becoming the Committee for Scientific and Technology Policy (STP), with a Working Party on Innovation and Technological Policy.

Why did governments espouse the policy objective of fostering technological innovation in the 1960s? There already existed many empirical studies concerning what was then called "science and technology". In addition, many policy measures on science and technology had already been implemented. But there was no integration of the results or knowledge (that is, the factors or conditions leading to economic progress), no overall framework, and no cohesive policy. As Charles Carter and Bruce Williams put it, in the late 1950s "technological innovation" did the job. The term allowed policymakers to discuss a series of activities in a new and inclusive way: coupling (a key word at the time) supply with demand, considering social as well and economic objectives, and integrating diverse policies (research, education, labor, industry). Technological innovation is a *total* process, a holistic system. Such a process or system includes everyone, but is oriented to the market, and this fact justifies orienting the research system to the market too.

6. Inventing innovation policy

The representation of innovation as a process and a system led to inclusive or integrative policy. An innovation policy is inclusive, it is a whole (or a "policy mix" as some called it) (Borras and Edquist, 2013; Flanagan et al., 2011). It concerns several ministries and public agencies; it coordinates many measures or policy instruments; it addresses a large range of issues. In terms of government action, whole or total means the "co-ordinated and concerted action" of several ministries (OECD, 1966: 7) and the combination of direct (funding) and indirect (climate) measures (p. 10).

Policy as obstacle to innovation

The Arthur D. Little Inc. study of 1963 for the US National Science Foundation (NSF) was the first of two studies commissioned from the consulting firm. In 1972, the NSF awarded another contract to the firm, conducted in cooperation with the Industrial Research Institute (Arthur D. Little Inc., 1973). Written by Michael Michaelis, from Arthur D. Little Inc. and contributor to the OECD report of 1966, and William Carey, and conducted between August 1972 and January 1973, the study looked at "obstacles" to technological innovation and suggested policy options for overcoming them. To the authors of the report, the obstacles are institutional rather than technical: market, finance, human resources and government regulations.

To Michaelis and Carey, "technological innovation is the process by which an idea or invention is translated into the economy ... the crux of the innovation process is utilization" rather than research (Arthur D. Little Inc., 1973: 1–2). *Barriers to Innovation in Industry* recommends that (1) policies related to technological innovation (economic, environment, taxation, trade, and so on) be better coordinated: there is need for a "focal point" in the executive branch of the federal government; (2) policy objectives should be clearly defined: there is a "zone of confusion" between "advancing the search for new scientific and technical knowledge (research and development) and stimulating the application of this knowledge (innovation). Policies that are efficacious for research and development will not necessarily trigger or excite the process that we term innovation"; (3) policies should focus on creating a market demand to meet public needs (adopting the *pull mode* rather than the *push mode*); (4) policies should be sector-specific, rather than "universal solutions", and require interactions between business and government.

Already in 1964, in an interview in *International Science and Technology*, Michaelis talked of our attitude toward, or popular faith in, miracles out of research and development. "The problem is not a matter of science and technology alone. Rather it is one of *institutional obstacles to the use* of technical innovation", that is, "the commercial use of modern technology", these obstacles being legal, regulatory, financial, labor-related, managerial (Michaelis, 1964: 40). We should start thinking in terms of systems, not products (for example, Bell Laboratories, Department of Defense). "The challenge to industry and government . . . is to organize not only the search for knowledge through research and development as we are already doing, but – equally vigorously – the use of that knowledge" (Michaelis, 1964: 45). "The basic requisite for the systems approach is active collaboration of all organizations – industry, government, and labor – that have a vested interest in a common objective" (p. 46). There is need for a Council for Science, Engineering and Policy, and industry should take the initiative to create such a council (p. 46). This council would not be a technical council but rather a managerial one.

Policy for industrial innovation

The idea that some public policies, particularly regulations, are a hindrance or "obstruction" to innovation – terms used respectively by the US Chamber of Commerce in 1966, the US Department of Commerce in 1967, and Keith Pavitt in 1963 – continues to prevail in many circles today. For example, the OECD Oslo Manual on surveying industrial innovation asks questions about government barriers to innovation (OECD, 2018b: 160). But to many others, innovation policy is a "stimulus" to technological innovation. To policymakers, innovation bears fruit, and the expected social payoff is a sufficient reason to support innovation. Innovation should be supported publicly because it is good in itself.[1]

The OECD

Policy as an inclusive notion has a long history at the OECD. This has applied to science policy and science and technology policy successively (see Appendix 1). But innovation policy is perhaps the most inclusive policy.

"Each country must develop its own policy" stated *Government and Technological Innovation* in 1966. The document was the result of a huge survey of policies among OECD Member countries. In fact, the bulk of the report is concerned with "government measures to stimulate innovation". In July 1964, the Committee on the Economics of Science and Technology sent three questionnaires to 12 governments on "the role of government in stimulating technical innovation" (OECD, 1965). The OECD asked for information on: (1) "research and development work [undertaken] to develop products or processes"; (2) the transfer ("spin-off") of military and space research to the civilian economy; (3) other measures such as cooperative research institutes, technical information and advisory services, government buying procedures (procurement). The first two activities are what came to be called supply- and demand-oriented policies. From this survey, one observes that the OECD's representation of inno- vation is clearly research-oriented, but also that the organization includes a whole set of measures. From a study of the questions sent to countries it is apparent that, at the time, the OECD had a clear sense of what an innovation policy is or should be.

Innovation policy as a concept is a construct. It combines existing (and ideal) measures – and would soon include new ones. According to *Government and Technological Innovation*, govern- ments "take action in areas which influence the conditions for innovation in firm [but] have no explicit policies to stimulate technical innova- tion" (OECD, 1966: 10). To the OECD, a "national policy" for technical innovation covers a large set of measures, as Charles Carter and Bruce William suggested in the late 1950s, and as the OECD sug- gested in the case of science policy in 1963. The majority of national policies reported by Member countries were demand-oriented. Supply-side or direct measures to industries (grants) is a practice that Member countries were not widely involved, except in a "tentative and experimental" way. Supply-side policy as an integral part of innovation policy is a later invention. The report discusses: (1) "development contracts" in strategic sectors, namely public money to firms for specific research and development projects, but mainly development, in order to develop new products for the public market and, to a limited extent, to strategic sectors deemed important for "economic expansion"; the report also considers public buying practices and procedures; (2) the use of military and space technologies for civilian purposes ("spin-off" or transfer); (3) other forms of support such as cooperative research institutes, technical information services, with a short section on tax credits.

The OECD's conclusion is that "the efforts made by Member Governments to stimulate technical innovation [apart from those concerned with the needs of the public sector are] relatively small" (p. 55). In the future, governments "should play a more active role" and develop "coherent policies" for technical innovation, rather than indirect and uncoordinated policies (p. 8), a discourse reproduced by scholars in subsequent years (for example: Emmanuel Mesthene, Chris Freeman).

Following the fourth ministerial meeting in October 1971, the OECD started a whole series of research programs on innovation and innovation policy (see Table 6.1). The first project was on innovation policy in "social sectors" or public services. This research program had the highest priority at the time. In Chapter 7, I discuss the study produced. The other project was on policy for industrial innovation, which over time acquired the highest priority on the OECD agenda. The ad hoc group on industrial innovation discussed in Chapter 5 collected information on 260 policy instruments or measures of technological

Table 6.1:   OECD projects on technological innovation

1972–86	Innovation policies
1973–6	Innovation in services
1978–81	Innovation in small and medium-sized enterprises (SMEs)
1981–9	Evaluation and impact of government measures
1986–	Reviews of innovation policies (France, Ireland, Spain, Yugoslavia and Western Canada)
1994–8	Best practices in technology policies
1994–2001	National systems of innovation (NSI)
1996–7	Technology diffusion

innovation, so called, in 18 Member countries.[2] After six years of work, the group produced a report in two volumes. To the OECD, policy for technological innovation is called "PSI (policies for stimulating industrial innovation)", comprising measures from research policy to industrial policy. PSI "are sets of measures instituting new provisions or modifying existing ones so as to . . . promote scientific knowledge for the purposes of innovation [and] the application of this knowledge to new technical achievements or their dissemination" (OECD, 1978: 28). The OECD typology covers three types of instruments or measures called specific, non-specific and major programs (see Appendix 2).[3] The measures embrace science policy (research and development), industrial policy (patents, standards and taxation) and other policies such as education and trade.

PSI is an inclusive notion. As Chris Freeman put it in the Preface to the analytical volume: "More than any other government policies, PSI represents that area of decision-making where science policy, technology policy and industrial policy converge and become almost indistinguishable" (OECD, 1978: 7). Yet the report noted that research and development remains, to governments, the ultimate policy instrument of technological innovation. Governments try to stimulate all the phases of the innovation process but "focus their main efforts on the first steps of the process, especially on research and development" (pp. 18–19).

The OECD project on innovation policy was followed by many others, such as a study of innovation policy for small and medium-sized companies (OECD, 1982). This continues today, with the reviews of innovation policy in Member countries – and has been imitated at the United Nations Conference on Trade and Development (UNCTAD), beginning in 1999.[4]

The Science Policy Research Unit

In 1966, economist Chris Freeman created the Science Policy Research Unit (SPRU) at University of Sussex, Brighton as a spin-off from his experience at OECD. In the fall of 1973, four governments (France, Germany, the Netherlands and the United Kingdom) commissioned the SPRU to conduct a study "to develop a framework . . . to be of assistance to government policymakers towards technological innovation in industry" (SPRU, 1974: 10). The exercise was similar to that of the OECD on PSI and others in the United States (see below). The objective was to give shape and life to a new concept: innovation policies. The report, prepared by Keith Pavitt and W. Walker, was submitted a year later (November 1974), then transformed into an article published in *Research Policy* (Pavitt and Walker, 1976).

"Until recently" states the report, "very little attention has been given by economists and members of other academic disciplines to the subject of government policy towards industrial innovation: government praxis is ahead of any theory" (SPRU, 1974: 12). "Government policies towards industrial innovation are often not clearly distinguishable and defined" (p. 59). The SPRU framework consists of a policy mix in the form of a threefold typology of policy instruments for innovation – direct, indirect and others (see Appendix 3). Research and development-related measures hold a large place in the narrative of the authors, despite a definition of innovation as a process composed of many activities.[5]

The US Department of Commerce

In 1971–2, President Nixon reoriented science and technology policy towards national and social needs (environment, energy, transportation) and innovation (commercialization).[6] "The mere act of scientific discovery is not enough" claimed

Nixon in the first ever Science and Technology Presidential Message to Congress in March 1972. "Even the most important breakthrough will have little impact on our lives unless it is put to use ... we must combine the genius of invention with the skills of entrepreneurship, management, marketing and finance". Nixon offered an "overall strategic approach" to policy in six points:

- Strengthening the federal role.
- Supporting research and development in the private sector.
- Applying government-sponsored technologies.
- Improving the climate for innovation.
- Stronger federal, state and local partnerships.
- Partnership in international science and technology.

It is in this context that the Department of Commerce was encouraged to "appraise, on a continuing basis, the technological strengths and weaknesses of American industry". In the fall of 1971, the department initiated a "study to identify the principal enhancement programs underway in a number of advanced countries: Canada, France, Japan, the United Kingdom and Germany". *Technology Enhancement Programs in Five Foreign Countries*, published in December 1972, is the analytical report that came out of the study (US Department of Commerce, 1972). "The United States is perhaps the only advanced nation in the world which has not undertaken national programs to stimulate technology development in the civilian sector" (p. 7). The reasons identified were: ideology and doubt about the effectiveness of such programs, inexperience with programs and lack of adequate data and information. The department study collected information on three aspects of policy: government involvement, organizational structure and government–private sector relationships.

The study has several limitations, the most important one, for analytical purposes, being the fact that national policies are not analysed and compared in a synthetic manner. Most of the document (288 pages out of a total of 328) is composed of appendices by countries. Yet, the study served as a starting point to a subsequent study.

The Center for Policy Alternatives

In 1972, Herbert Hollomon created the Center for Policy Alternatives at MIT. At about the same time as SPRU, namely between 1973 and 1975, the center conducted a study of science and technology policies in five countries (France, West Germany, the Netherlands, the United Kingdom and Japan) (Center for Policy Alternatives, 1975). Hollomon and Thomas Allen acted as principal investigators. Among the contributors were William Abernathy and James Utterback – the latter was an assistant on the National Planning Association study conducted for the NSF. As SPRU had, the results of the study were published afterwards in *Research Policy* (Allen et al., 1978). Hollomon himself published articles on the results of the study in the following years (see Utterback et al., 1976).

The final report starts by noting the "absence of explicit 'national innovation policy'" to inform the US government on the "need for coordinated science and technology policy" (Center for Policy Alternatives, 1975: I–2) and "Aspects of the innovation process which can be manipulated by decision-makers" (pp. I–13). The study is based on a framework or model of the "life-cycle" of innovation. Unlike that of the SPRU, the report starts with firms' needs, not the problems of policy. "Firms which are at different positions in the evolution of their product and process technology will respond to differing stimuli and undertake different types of innovation" (pp. I–24). As an innovation matures, it shifts from product to process innovation,[7] from a multitude of products to a standard form. In the following years, Utterback developed the idea in the academic literature and built a reputation on this framework.

What does the "life-cycle" mean for policy? "Different government measures may be effective in influencing technological innovation at the different stages" of the innovation cycle (pp. I–29). "Government efforts [must be] specifically designed to affect the innovation process" differently, depending on the lifecycle of products/ processes (pp. I–3). The center developed a

typology composed of 12 categories of measures
(mechanisms) "influencing" (a term used by the
US Department of Commerce, 1967) or "affecting"
innovation, grouped under three types of instruments (vehicles): the innovation process within the
firm, the intellectual base, the social consequences
(see Appendix 4). Influencing and affecting are the
appropriate terms. Innovation policy is a mix of
measures that affect or influence innovation, and
that were not historically or originally conceived
for this end:

> Governments may be able to influence the rate and direc-
> tion of technological change through their ability to alter
> industrial development by affecting market forces, the
> structure of rewards, industrial organization, and the
> regulatory constraints within which industry operates.
> Governments can also alter the resources available to
> firms – informational, financial and human resources –
> and the allocation of these resources. In affecting either
> the environment, or the resources available to industrial
> firms, governments may help to initiate technological
> change, to sustain change, or to regulate change (pp. I–6).
> [However, governments do not have Departments of
> innovation, but a "multitude" of measures (pp. I–8)]. A
> great variety of US government programs have an impor-
> tant relationship to technological innovation. Many,
> if not most, were not intended to affect innovation . . .
> It is useful for conceptual purposes to organize these
> government programs and activities into a framework.
> (Hollomon, 1980: 31–2)

What does innovation mean to the Center for
Policy Alternatives? "The term innovation, as
we will use it, refers here to technology actually
brought by the firm into first commercial use or

application ... The process ... occurs in three phases: idea-generation, problem-solving or development, and implementation and diffusion" (Center for Policy Alternatives, 1975: I–12). Three key terms in this representation of technological innovation are: originality (first) that goes back to the 1960s,[8] commercialization and process.

The center conducted a further study for the "Government Policies for Innovation Project" of the Office of Technology Assessment (OTA) in 1978 (Center for Policy Alternatives, 1978; summarized in Hollomon, 1979). At OTA, the project manager was Ellis Mottur.[9]

Development of the idea

Research on innovation policy in the United States developed in the following years supported by, among others, national agencies and research councils. For example, the NSF's National R&D Assessment Program, established in 1972, funded many academic studies on innovation policy, among others (for example, Roger Noll at the California Institute of Technology, Noll, 1975; Christopher Hill and James Utterback, 1979). The organization itself initiated the series with:

- 1973: National Science Foundation, *Serving Social Objectives via Technological Innovation: Possible Near-Term Federal Policy Options*, Office of National R&D Assessment.

- 1973:   National   Science   Foundation, *Technological Innovation and Federal Government Policy*, Office of National R&D Assessment.

Private firms also benefited from the National R&D Assessment Program. Several studies were commissioned from consultants (for example, Practical Concepts Inc. – Posner and Rosenberg, 1974; Harbridge House Inc., 1975; Gellman Research Associates, 1974, 1976). In the 1980s and 1990s, Louis Tornatzky, the administrator of the program for some years, summarized the results arising out of the research program and produced syntheses of what was known about innovation (Tornatzky et al., 1983; Tornatzky and Fleisher, 1990).

Typologies of innovation policy also multiplied. Beginning in the 1970s, the US Congressional Research Service (CRS) conducted many reviews of the existing literature on innovation (for example, US Joint Economic Committee, 1975). One of these, from the pen of Mary Mogee, brings in a further typology of policies. As part of a Special Study on Economic Change conducted between 1978 and 1980 for the Joint Economic Committee, the CRS produced a staff study on innovation (called Research and Innovation Area Study), as Volume 3 of the main report, accompanied by eight technical papers (US Joint Economic Committee, 1980).[10] The study analysed 42 innovation studies, as it called them, produced over the last two decades, and identified 205 policy recommendations made to "stimulate innovation", classified into 14 policy areas (pp. 140–48) (see Appendix 5).[11]

Measuring innovation

Statistics helped to give life to the concept of innovation policy, as the statistics on research and development had for research policy (Godin, 2005). The NSF and the OECD developed methodologies for collecting data on research and development in the late 1950s and early 1960s. Early on, the OECD Frascati Manual for measuring research and development was envisaged as a contribution to "a systematic study of the relationship between scientific research, innovation and economic growth" (OECD, 1962: 19). However, innovation had to wait until later for measurement, leaving full place to research and development to be used as an indicator or "proxy" for innovation for decades to come.

Measurement of innovation started in the 1970s at the official level. At the time, the surveys were experimental. Using the National R&D Assessment Program, the NSF funded four studies in the mid-1970s to assess the feasibility of measuring innovation (Fabricant et al., 1975; Hildred and Bengston, 1974; Posner and Rosenberg, 1974; Roberts and Romine, 1974). Every study contained doubts about measuring a fluid concept such as innovation. Yet indicators were soon produced. Based on a contract awarded in 1974 to Aaron Gellman's consultancy firm, Gellman

Research Associates, the NSF produced its first compendium of indicators on technological innovation in the 1970s (Gellman Research Associates, 1974, 1976), soon to be followed by some statistics included in *Science and Engineering Indicators*, a series that still exists today. In the 1980s, the NSF's National R&D Assessment Program continued to fund feasibility studies on indicators (see Hansen et al., 1984) and the National Science Foundation conducted its first innovation survey in 1986.[12] The Center for Policy Alternatives conducted its own exercise too (Hill et al., 1982; Hill et al., 1983), as did the US Congress (US Committee on Ways and Means, 1980).

By the late 1980s, most countries had conducted at least one official survey of industrial innovation (see OECD, 1990), some going back to 1970, like Canada's. Building on these sources, the OECD produced a methodological manual (the Oslo Manual) containing guidelines for measuring technological innovation in firms, using the system approach as framework (OECD, 1991). In the latest edition of the Oslo Manual (2018), this framework changed to the System of National Accounts, as the OECD Frascati Manual has since 1963. In the next two decades, the Oslo Manual was extended to cover innovation in services, organizational innovation and design innovation in addition to technological innovation.

During the same interval, international organizations began producing compendiums of indicators. The OECD started its own in 1995: the

Science, Technology and Industry Scoreboard. The scoreboard is published every other year, alternately with the *Science, Technology and Innovation Outlook.* The European Union also started its own scoreboards: the *Innobarometer* (first edition: 2001), the *European Innovation Scoreboard* (2001) and the *Regional Innovation Scoreboard* (2006).

Notes

1. Market failures are not among the original causes of policy. The argument on market failures is an ex post rationalization coming from economists (see Averch, 1985).
2. Technological innovation "is defined as comprising those technical, industrial, commercial or other steps which lead to the successful marketing of new manufactured products and/or to the commercial use of technically new processes or equipment" (OECD, 1978: 25).
3. A later study had a similar typology of innovation policies for small and medium-sized companies (OECD, 1982: 151) as follows: public funding to industry (for example, research and development, tax credits, venture capital), services (information), economic policies (competition, procurement, patents, regulations).
4. See https://unctad.org/en/pages/publications/Scien ce,-Technology-and-Innovation-Policy-Reviews-(STIP-Reviews).aspx.
5. The report defines "technological innovation as the technical, industrial and commercial steps which lead to the marketing of new and improved manufactured products and to the commercial use of new and improved production processes and equipment" (SPRU, 1974: 17). The report contrasts this definition to "technical progress", defined as "increases in the productivity of labour, capital and land" or technological

change (the *Research Policy* article uses "technological progress" in this definition; "technical progress" is defined as "the diffusion of innovations" (Pavitt and Walker, 1976: 20).

6. The terms "innovation" and "technological innovation" appear regularly in Nixon's message to Congress.

7. And from radical to incremental product.

8. Mansfield, 1961; OECD, 1966, 1968; Pavitt and Wald, 1971. A few uses before the 1960s are from sociologist Colum Gilfillan and economist historian Rupert Maclaurin.

9. A "successor" to Hollomon's work was that of Lewis Branscomb, American physicist, government policy adviser and corporate research manager, and a prolific writer on innovation policy (Branscomb and Keller, 1998).

10. For example, Richard Nelson on basic research as an important part of an effective policy package; Albert Rubenstein on incremental innovation.

11. The Congress defines innovation as "the process by which society generates and uses new products and manufacturing processes. It includes the activities ranging from the conception of an idea to its widespread application by society" (US Joint Economic Committee, 1980: 11). "Although the contribution [of basic research to economic growth] is usually indirect and delayed, science seems to act as an 'engine' of technology, without which specific innovations may be delayed or prevented and technological progress may eventually stagnate" (p. 264).

12. *A Survey of Industrial Innovation in the United States: Final Report*, Princeton, NY: Audits and Surveys – Government Research Division, 1987; NSF (1987), *Science and Engineering Indicators 1987*, Washington, pp. 116–19.

7. Innovation policy today

By the early 1980s, innovation had emerged "as a national hot topic" (US Joint Economic Committee, 1980: 2). Now that the idea of technological innovation had been constructed (as explored in Part I) and the concept of innovation policy too (as explored in Chapter 6), it was time to act. When did innovation policy begin? If one consults the OECD report on policies for stimulating innovation of 1978, innovation "measures" existed since the 1930s (OECD, 1978: 56). However, if one starts with the term innovation as such, innovation "policies" began in the 1980s.

Why such a difference in time, between measures in the 1930s and polices in the 1980s? One reason is the "fuzziness", as the OECD calls it, of the concept of innovation policy. One review of the time is conscious of the fuzziness: "depending upon one's definition of innovation, there were between 30 and 200 such bills before the 96th Congress" (1980): tax incentives, productivity, small business, patents, research and development (US Joint Economic Committee, 1980: 6). But these were separate bills without interrelationships, without "a clear signal" as to the direction and priorities of national innovation activities" (p. 9). Another reason is the inclusive definition of innovation. As an OECD

Overview of Public Policies to Support Innovation put it in 2005: "almost every kind of public policy has either a direct or indirect impact on factors that affect innovation activity" (OECD, 2005a: 4). As a consequence, one can read back in time and find *de facto* innovation policy everywhere in the past. For example, since research and development is part of the policy mix, innovation policy could be considered to have begun after the Second World War. Be that as it may, before the 1980s, innovation policy was informal or indirect; a so-called series of measures with no overarching design. An explicit policy brings order and implements the measures toward a specific goal. A "national strategy" does even more. It orients or aligns a series of policies specifically toward innovation.

Notwithstanding the Norwegian "Education Innovation Act" of 1954, the first innovation policy carrying the concept of innovation as such is from 1980. Following a review of innovation policy conducted in 1978 and 1979 by the US Advisory Committee on Industrial Innovation, chaired by the Secretary of Commerce, President Jimmy Carter announced an "Industrial Innovation Initiative" on 31 October 1979, composed of 31 "actions" in nine areas "that will ensure our country's continued role as the world leader in industrial innovation" (Office of the White House Press Secretary, 31 October 1979):

- Enhancing the transfer of information.
- Increasing technical knowledge.

- Strengthening the patent system.
- Clarifying anti-trust policy.
- Fostering the development of small innovative firms.
- Opening federal procurement for innovations.
- Improving our regulatory system.
- Facilitating labor/management adjustment to technical change.
- Maintaining a supportive climate for innovation.

For several years, there were concerns in the United States that the country's lead in innovation had declined relative to the OECD, particularly when compared to Japan (US Joint Economic Committee, 1975). As Herbert Hollomon (1980) put it:

> Most countries have been ahead of the United States in their concern for industrial innovation and industrial technology with respect to economic development. The discussions we are now having in this country began in Europe and Japan several years ago. (Hollomon, 1980: 197)

> There is ... resistance in the private sector to such support, perpetuated by the myth that most firms and most product development occurred in this country without government interference. (p. 203)

> [But there is a danger coming:] Most major technological developments will take place outside of the United States in the future. (p. 197)

"There is no one place where the Federal government can take action", claimed Carter, "we have therefore chosen a range of initiatives." To the Carter administration, "innovation is a subtle and intricate process, covering that range of events from the inspiration of the inventor to the marketing strategy of the eventual producer." A week later, an editorial in *Nature* called the initiative a "package" of measures: it "provides a central theme around which otherwise disparate activities can be arranged . . . The problem, however, is that innovation itself is a slippery concept that is almost impossible to define" (Nature, 1979).

There is little word on research in the initiative, except for support to university–industry research and development projects. The "problem" with universities is that they "are far removed from useful and productive product and process embodiments", claimed the Advisory Committee (US Advisory Committee on Industrial Innovation, 1979: 18). "It is not enough for the inventor to invent; he must also bring his idea for a new product or process to the market" (p. 6). As one Senator put it, on the day of Carter's announcement of the initiative, "what counts . . . is how effective a nation is in transforming science into commercially marketable products. This is what innovation policy is about" (US House of Representatives, 1980: 3).[1]

A year later, the US Congress passed the Technology Innovation Act (21 October 1980). To the Carter administration, a central solution was the commercialization of research through Centers for Industrial Technology. Offices of Research and Technology Applications were promoted in every federal laboratory and university. The Act gave rise to the famous Bayh-Dole Act of December 1980.

This was a key moment in the shift from research to innovation considerations in policy. Over the next decades, the commercialization of university research became a priority of public policy. In every western country, council grants to universities are increasingly geared toward users. Research policy is mixed, to different degrees, with innovation policy.[2] In most countries, science departments are merged with economic departments. Departments or administrative units in charge of research policy now define their fields of responsibility as "research and innovation". The EU Commission, for example, has established a Directorate-General for Research and Innovation (formerly DG Research). EU member states have meanwhile adopted the label "research and innovation" for funding programs and directives. In the UK, the government set up the UK-Innovation Research Center in 2018, merging research and innovation in a single agency. These moves are a consequence of the discourse on the marginalization of research, which began in the 1960s ("research is not enough").

Innovation strategies

What soon became popular is the concept of a "national strategy". The policy mix finally got a name: strategy. The concept was on everyone's lips in the 1980s.[3] The term is borrowed from the military field. An innovation strategy is a plan, with a framework that serves to organize, or rather rationalize, policy measures. A strategy is a series of objectives to be embraced by every department of the administration and all organizations in society. It is an agenda for effective policy. It orients existing measures towards innovation and suggests avenues for coordinating the measures and initiating new ones.

By now, most OECD countries (33 out of 35, according to the OECD) have national strategies for innovation (with a research and development target of 3 per cent of gross domestic product) (GDP). The great majority of these originated in the 2000s and 2010s (OECD, 2018a: 26). In fact, European countries must have such a strategy, according to the European Commission. Furthermore, 23 European countries have regional strategies.

One of the first such strategies was the *Strategy for American Innovation* led by Barak Obama in 2009, and updated in 2011 then in 2015, and followed by a White House Office of American Innovation in 2017. Soon the OECD developed its own strategy, modeled on the American experience (a plan plus new measurements). In May 2007, OECD ministers mandated the preparation of an innovation strategy, a *"sine qua non* of economic growth and productivity" (OECD, 2007). What was required was a "co-ordinated, coherent, 'whole-of-government' approach", a "cross-disciplinary mutually-reinforcing package" (p. 27), a "comprehensive cross-government policy approach" (OECD, 2008a), an "ecosystem view" (OECD, 2008b). The strategy was prepared for the ministerial meeting of 2010 (26–8 May) and included the following documents:

- *The OECD Innovation Strategy: Getting a Head Start on Tomorrow*
- *Measuring Innovation: A New Perspective*
- *Innovation and the Development Agenda*

The strategy is a mix of policy instruments ("a broad concept of innovation", OECD, 2010: 15, 34–8; a match between supply-side and demand-side instruments). The oldest typology of policies, by type of measures or instruments, is replaced by a typology of challenges called "priorities":

- Empowering people to innovation (education and skills, international mobility).
- Unleashing innovation (economic policy, regulations and administrative burdens, labor policy; public support: tax provisions, venture capital, credit, subsidies).
- Creating and applying knowledge (public research, property rights, open information).
- Addressing global and social challenges such as climate change (scientific cooperation and technology transfer).

- Governance and measurement (make innovation a central component of government policy; "a whole-of-government approach").

The strategy is relatively immune to or does not espouse explicitly the discourse on the marginalization of research. The "mythology", as practitioners would call it, of research as source of technological innovation is strong. For example, according to a recent report from the National Science Foundation, research and development is (still) a major driver of innovation (US National Science Board, 2016). Certainly, echoing a review of the OECD work on innovation,[4] the OECD strategy suggests that "policy will need to move beyond supply-side policies focused on research and development, and specific technologies, to a more systemic approach that takes account of many factors and actors" (OECD, 2010: 11). At the same time, the OECD also maintains that "basic research in particular, provides the seeds for future innovation" (p. 10), and that "science continues to be at the heart of innovation" (p. 13). Nevertheless, and unlike the past, the series of measures on research and development does not appear first in the list of priorities.

The OECD strategy was updated in 2015 (ministerial meeting, 3–4 June) as *The Innovation Imperative*, a "toolbox for governments that wish to strengthen innovation and make it more supportive of inclusive and green growth" (OECD, 2015: 3). The vocabulary on challenges or priorities for 2010 shifted to "agendas":

- Boosting productivity growth.
- Closing the development gap of low- and middle-income countries.
- Diversifying the economy and moving up the value chain.
- Innovation in natural resource-based economies.
- Innovation for inclusive and green growth.

The OECD is not the only international organization to have developed an innovation strategy. In the case of the European Union, one may go back to the *First Action Plan* (1996), then the *Lisbon Strategy*, the *Innovation Initiative* (updated in 2006) and the *Innovation Union Flagship* (2010). These strategies emerged out of a series of previous policies, as the European Commission's historiography puts it (European Parliament, 2016): from research policy (*A Policy for Research and Innovation in the Community*, 1967) to industrial policy (*A Policy for Industrial Innovation*, 1981)[5] to a policy mix (*Green Paper on Innovation*, 1995).

Broadening the concept

Along with strategies, innovation came to broaden its meaning from a strictly technological matter to an all-encompassing notion. In the 1980s and 1990s, a series of new terms that I call "X-innovation" appeared, competing with industrial innovation or innovation coming from industry, and that continue to contest technological innovation as a

Table 7.1:   X-innovation

Oldest (an object)	Newest (an adjective/a
Technological innovation*	metaphor)
Product/process innovation	Inclusive innovation
Industrial innovation	User innovation
Marketing innovation	Free innovation
Organizational innovation*	Democratic innovation
Educational innovation	Common innovation
Political innovation	Open innovation
Social innovation*	Digital innovation
	Hidden innovation
	Disruptive innovation
	Reverse innovation
	Frugal innovation
	Jugaad innovation
	Responsible innovation
	Sustainable innovation
	Grassroots innovation
	Eco-innovation

Note: * Another word used in place of "innovation" in these terms is "change".

hegemonic discourse: social innovation, sustainable innovation, responsible innovation, and so on (Godin, 2019). To make sense of this linguistic innovation, it is useful to distinguish an X-innovation according to the date of its appearance (see Table 7.1). People began theorizing about X-innovation in the 1950s. X-innovation was then concerned with an object, such as technology, industry, organization or education. In a second step, during the 1980s and 1990s, new forms appeared that defined innovation using adjectives (Gaglio, 2012). Certainly, adjectives existed since the 1930s in the case of typologies of technological innovation. Innovation is so banal (that is, so frequently occurring or commonplace) that scholars need to qualify or redefine and redescribe innovation. Innovations are classified according to their effects on the economy or society: (1) major, revolutionary, radical, paradigmatic, systemic; (2) minor, incremental. For several decades, those who studied industrial innovation did so by limiting innovation to "major" innovation, under different terms (radical, revolutionary, disruptive). Major innovations are those with structural and systemic effects. Other types of innovation are relegated to a second-order level, being incremental or gradual ("minor" innovation).[6]

In contrast, recent types of innovation consider modes of innovation. Now an adjective, rather than an object, defines what innovation is; it can be social, responsible, sustainable. This has to do with the "quality" of innovation: we need a different type of innovation, so it is claimed. Let us stress two characteristics of the newest X-innovation forms. Firstly, the societal. On one hand, that is, the input or process side, X-innovation emphasizes inclusion, namely the participation of diverse publics in deliberations about innovation from an early stage in decision processes. Hence, we have X-innovation forms such as inclusive innovation, democratic innovation and free innovation. On the other hand, on the outcome side, X-innovation places the emphasis on societal, ethical and environmental considerations. There is a moral imperative here. Innovation must be social, responsible and sustainable.

Educational innovation

One of the first X-innovations in public documents is "educational innovation". Early on in the history of innovation policy, "scientists and engineers" as a category of human resources in research and development gave rise to the issue of education and training for innovation. Scientists and engineers must be better equipped to deal with technological innovation issues. The concept of and discourse on educational innovation emerged in this context. Most of the documents on educational innovation produced in the 1960s deal with education at all levels. But some are concerned with engineers specifically.

The President's Science Advisory Committee

In 1961, the President's Science Advisory Committee (PSAC) established a *Panel on Educational R&D*. The Panel reports to the US Commissioner of Education, the Director of the National Science Foundation (NSF), and the President's Special Assistant for Science and Technology, who was also the Chairman of the PSAC. The panel's principal interest was the promotion of educational research and development to improve the quality of education. The panel was chaired by Jerrold Zacharias (Professor of Physics at MIT and member of PSAC). Three years after the PSAC request, the panel published a report, *Innovation and Experiments in Education*, dealing with new materials, methods and curricula and new school systems (US President's Science Advisory Committee, 1964). The report was the result of a series of seminars and two conferences held between November 1962 and June 1963, financed by the Office of Education and the National Science Foundation.

The report looks at three areas of reform or innovation in elementary and secondary education – curriculum development and materials, teacher education, and organizational or "institutional innovation" (for example, classrooms, teaching hours, relations with the community) – and suggests a unit of experimental schools to test the proposals and serve as model for the education "system". This was the subject of many studies on educational innovation at the time: the study of the whole system, at every level of education.

The Department of Health, Education and Welfare
The most productive agency on educational innovation in the United States in the 1960s and 1970s was the Office of Education of the Department of Health, Education and Welfare. The office commissioned many studies. Most of them had the same focus as PSAC. The studies looked at primary and secondary levels of education. The first such study was commissioned in March 1964. The office commissioned a study from System Development Corporation to test the Department

of Agriculture's "field extension service concept" to education: site visits of educators to innovative schools to disseminate innovation practices and reduce the "time lag" or "gap" between innovation and practice (Caffrey, 1965; Richland, 1965).[7]

Another contractor conducted a survey among science opinion leaders and schoolteachers at the primary level to evaluate the role of the former in the adoption of innovation in teaching (Mechling, 1969). Still another study commissioned from Arthur D. Little Inc., in which innovation is "a deliberate, novel, specific change" in educational practices, developed a "response to a need model" of the innovation process, with the aim of increasing the speed of adoption of innovation in public schools. The consultant firm applied the model to organizations (school districts) as adopters of innovation – rather than individuals, as most models of the time did – and considered the adjustments or feedbacks necessary in the adoption process (Arthur D. Little Inc., 1968).

Other contractors looked at research on education and the application of results to practitioners' needs. Ronald Havelock, founder of the Center for Research on the Utilization of Scientific Knowledge at the University of Michigan, developed the first sketch of his "linkage model", as he called it, in 1967–8 under a contract from the National Center for Educational Communication, Office of Education, and tested the model in subsequent years (Havelock, 1969, 1970; Havelock and Havelock, 1973). Based on an extensive literature on existing models of innovation and a national survey of 500 school districts, Havelock offered the Office of Education a system view of innovation (from producers of knowledge to users), with needs, again, as the stimulus. To Havelock, an innovator is "the first person in a social system to take up a new idea" (Havelock, 1969: 7–13).

Such an approach continued to be popular in the 1970s at the Department of Health, Education and Welfare. In the fall of 1976, the National Institute of Education commissioned a study from the Center for the Interdisciplinary Study of Science and Technology at Northwestern University to explore ways to bring educational research and development results to practitioners, and to feed forward their needs and concerns to educational researchers and developers and their sponsors (Radnor et al., 1977: 8). Many researchers from different universities were involved in the project, including Ronald Havelock, Everett Rogers (Stanford University) and Gerald Zaltman (University of Pittsburgh). The result was the "Research and Development Exchange (RDx)" model, again inspired by the century-old agricultural extension program of the Department of Agriculture, and not dissimilar to Havelock's model.

The Department of Commerce
Innovation for higher education soon got the attention of governments. Technological innovation being considered a practical affair, engineers must be better equipped to this end, so it was said.

In 1966, Daniel de Simone, who contributed to the PSAC conference of 1964, organized a conference on engineers and education, with the theme of "creativity". The conference was sponsored by the National Academy of Engineering, the National Science Foundation and US Department of Commerce. The aim of the conference "was to examine the creative processes of invention and innovation, the opportunities for encouraging creative activities in the engineering schools, and the possibilities for developing and supporting creative engineering education" (Simone, 1968: vii). Innovation was defined as "bringing new concepts to reality" (p. viii), "the complex process of introducing ideas into use or practice" (p. 4). To Simone, creative engineering education is education for innovation. "We have underplayed the role of engineering" (p. 7). "Technological invention and innovation are the business of engineering" (p. 4), as John Duckworth, Herbert Hollomon and Jack Morton put it. "Engineering is a profession, an art of action and synthesis and not simply a body of knowledge. Its highest calling is to invent and innovate" (pp. 1–2). It is a matter now of developing "the inventive and innovative potential of engineering students" (p. 11):

> In the schools of engineering, the art of engineering has been largely neglected. The stress has been on [knowledge and] analysis rather than synthesis, on the abstract rather than the messy alternatives of the real world . . . The subjects have little to do with the use of technology to solve real social and industrial problems . . . Graduate

students, especially, should be instilled with this sense of mission . . . the progress of mankind . . . The best way to develop and foster inventive and innovative talents in engineering students is to involve them in projects and experiences that stimulate and require such talents. (pp. 12–14)

Herbert Hollomon's introductory contribution to the conference continues on the same lines. Engineering education is:

too science-based [science-oriented: learning more rather than learning how] . . . removed from the creative act that the engineer or inventor has to perform to bring the results of science and technology to the benefit of society . . . The limitations [to social problems] are social, political, and economic, rather than scientific . . . We need to bring together persons who, on the one hand, are able to understand the needs of the society in which they live, and, on the other hand, are sufficiently familiar with the techniques and capabilities of science to be able to invent solutions . . . These subjects should be essential elements of the education of engineers . . . The ultimate task of establishing new institutions that are required to bring the changes society needs must be given to the entrepreneur – the innovator . . . Engineering education has removed itself from a concern with the social, economic and political problems of our time and has become more concerned with the science that underlies and undergirds the engineering we practice . . . [We need to] "put science to work". (pp. 23–30)

The Center for Educational Research and Innovation

At about the same time as the American agencies, the OECD produced thoughts on innovation in education – without specific attention to the higher-education level at the beginning, at least as the concept of educational innovation was used in the 1960s and 1970s. The thoughts of the organization cover every level of the school system.

In June 1968, the organization created the Center for Educational Research and Innovation (CERI). CERI organized a conference in November 1968 (OECD, 1968a), followed by a workshop in July 1969 (OECD, 1969), then produced technical studies, later collected in a series of case studies (OECD, 1973). The organization urged the development of national and local educational "innovation policies", also called "strategies" (OECD, 1973) and of an "Educational Innovation Agency" in every Member country (OECD, 1969).

CERI defines innovation as "those attempts at change in an educational system which are consciously and purposefully directed with the aim of improving the present system": organizational, technological and curricular (p. 13). CERI is conscious of the fluidity of the concept of innovation:

> The term "innovation" has become a complex one taking on a variety of meanings in a variety of situations. It is difficult to determine or analyse the real meaning of it. It even seems to be a goal in itself. (p. 15)

Nevertheless, CERI uses the concept to define the educational system as an "innovative system" – this is one of the first uses of the term. An innovation system is open. It involves all relevant participants in decisions, including users, and it is experimental, flexible and self-renewing (pp. 19–20).

CERI continued to study educational innovation in the next three decades, then contributed to the OECD innovation strategy of 2010 and produced several related documents in the following years:

- *Measuring Innovation in Education: A New Perspective*, 2014.
- *Innovation Education and Education for Innovation: The Power of Digital and Skills*, 2016.
- *Schools at the Crossroads of Innovation in Cities and Regions*, 2017.

In September 2016, the OECD also organized a conference on what it calls the "Knowledge Triangle" to better "integrate" education, research and development, and innovation activities and enhance their impacts on innovation (OECD, 2016).

Public sector innovation

The OECD strategy of 2010 broadened the concept of innovation to the public sector. *The Innovation Imperative in the Public Sector* (2015) and *Fostering Innovation in the Public Sector* (2017) are two such documents. The OECD also created an Observatory of Public Sector Innovation.[8] Public sector innovation is also a concern to the European Commission.[9]

The concern for public sector innovation is not new at the OECD. In response to the conclusions of the fourth ministerial meeting of 1971 that asked for an extension of the work on innovation to the public sector to increase its contribution to science and technology, the Committee for Scientific and Technological Policy (CSTP) suggested a program of work on "innovation in the social sectors" or public services (OECD, 1972c). Given the complexity of the issue, the group suggested constructing a socioeconomic framework and conducting two pilot studies: health and education. In the end, the research studied health, community (including urban), development, communications and social security. Case studies were conducted in eight participating countries. The research program gave rise to an analytical report in 1976, *Innovation in Services*, published as *Policies for Innovation in the Service Sector: Identification and Structure of Relevant Factors* in 1977.

The initial proposal claims that in contrast to the manufacturing sector where needs are defined in terms of the market, social needs are not fully articulated and governments could create markets for social goods. It implies a "system approach" (OECD, 1972c: 9). Too often, failures result from the attempt to import the diffusion model of manufacturing to social sectors. There is need for "new concepts of managing the research, development and innovation process" (p. 10).

The project owes a lot to the new service economy. In this context, the OECD stresses the "service dilemma" as "limits [on public expenditures] versus increased demand" (OECD, 1976: 10).

To the OECD, "Public services may be described as those activities that communities recognize as serving basic societal functions as such" (p. 22). "Something different – in approach, technique, perspective – is needed ... The name given to this different answer is innovation" (p. 10). To the OECD, innovation is "either the modification of the existing or the introduction of the new in the delivery of services" (p. 215).

Science and technology are only one aspect of the "ways in which innovative measures can be utilized to enhance service delivery". Originally, to the OECD, innovation in social sectors meant the contribution of science and technology to public sectors. By 1978, the organization changed its mind – about a "subject of long and heated discussions". One must consider the "totality" of the picture (p. 4). "Much of the sentiment, and even enthusiasm for innovation as a technique for successfully grappling with service issues arises from the long line of successes which innovation in industry has yielded ... The successes of industrial innovation were soon seen as an ideal model" (p. 10). However, there are "fundamental differences between the industrial and the service sectors": market forces versus public welfare (p. 11). "In the goods-producing sectors, the final output of the economic activity is a physically tangible product. Services distinguish themselves from goods by their intangible character" (p. 21). Innovation in public services "carries considerably beyond the framework of science and technology".

It includes a "wider set of factors": organizational and social.[10]

As a consequence, the types of innovation studied are technologies, programs, procedures and organizations. The OECD also urges a system approach: "the need to consider a varied and broader set of factors forces consideration of means, goals and the environment as a totality" (p. 214) and the "participation of all those with particular interests that may be affected" (p. 216).

Social innovation

Another highly popular term of recent innovation strategies is social innovation. In 2009, the US government created an Office of Social Innovation and Civic Participation. The concept got into innovation strategies in the 2010s (for example, the Innovation Union Initiative of the European Commission, 2010), defined as "new ideas that meet social needs, create social relationships and form new collaborations".[11] To the OECD, "Social innovation seeks new answers to social problems": new services that improve the quality of life of individuals and communities; new labor market integration processes, new competencies, new jobs, and new forms of participation, as diverse elements that each contribute to improving the position of individuals in the workforce.[12]

Like public sector innovation, the term social innovation has an old history in policy circles.

In 1969, the OECD Secretary-General invited an ad hoc group chaired by Harvey Brooks, Dean of Engineering and Applied Physics at Harvard University, with Jean-Jacques Salomon as Secretary, to make a reassessment of science policy in the face of recent disenchantment with science. The group called itself the "Group on New Concepts of Science Policy" (OECD, 1971). The new concept or perspective is one that considers social needs more seriously. The report of the group is a call for socially oriented technology, called "social technology", and policy. Some historiography raises doubts as to whether the report had any real influence on policy.[13] True, economic growth continued to define the discourses on innovation in the following decades.

Social technology is only one of the new terms that occurred regularly on practitioners' lips in the 1960s and 1970s. Social innovation is another. Donald Schon created the Office of Social and Technical Innovation in 1966, a non-profit social research and development firm in the Boston area. The US Bureau of Labor Statistics has a few words on social innovation in a report on technological change from 1966. To the Bureau of Labor Statistics, social innovation is a solution "to cope with problems raised by advances in technology": improvement and training of young people, retraining and increasing the mobility of the labor force, provisions for maintenance of income of unemployed workers (US Bureau of Labor Statistics, 1966: 2). The US Department of Commerce's report *Technological Innovation: Its Environment and Management* of 1967 has a whole section on social innovation, defined as solutions to social problems addressed in public policy. Robert Mueller from Arthur D. Little makes use of the concept too in his book *The Innovation Ethic* (Mueller, 1971). According to the OECD report on innovation in public services, "Economic innovation may be looked upon as a subcategory of social innovation", because new products and processes have social implications (OECD, 1976: 9).

Etc., etc.

The latest and most all-encompassing term is inclusive innovation, as "initiatives that directly serve the welfare of lower-income and excluded groups". The term entered into the discourse on economic growth after the OECD strategy of 2010, starting with OECD conferences in 2011 (Rio de Janeiro) and New Delhi (2015), then was included in the innovation strategy of 2015. The former conference stressed the need for an "innovative approach" to innovation policy and industrial policy; one that is open; that gives place to experimentation; and that considers seriously the contributions of peripheral countries (OECD, 2014). Three documents followed:

- *All on Board: Making Inclusive Growth Happen*, 2015.
- *Innovation Policies for Inclusive Growth*, 2015.

- *Making Innovation Benefit All: Policies for Inclusive Growth*, 2017.

This is not the end of the story. To this series of X-innovation forms, one may add sustainable (or green) innovation and responsible innovation. The first got into the OECD innovation strategy in 2015. The latter is a term that the European Commission adopted in its eighth Framework Programme of 2014 (Horizon 2020). The European Commission defines responsible [research and] innovation as "an approach that anticipates and assesses potential implications and societal expectations with regard to research and innovation, with the aim to foster the design of inclusive and sustainable research and innovation".[14]

To policymakers, X-innovation is an extension to the concept of innovation. It serves to give still more space (and legitimacy) to innovation. The concept is applied to different phenomena and fields. This is only one function of the newest X-innovation terms. In the academic literature, X-innovation is a contestation of, and an alternative to, technological innovation (Godin, 2019). For example, social innovation is regularly contrasted to technological innovation, and presented as a remedy for or adjustment to the undesired or limited effects of industrial and technological innovation – although the term social innovation appeared a century before technological innovation. The story is one of appropriation and contestation. On one hand, people appropriate a word (innovation) for its value-laden quality and, consequently, because of what they can do with it. A word with such a polysemy as innovation is a multi-purpose word. It works in the public mind (imaginaries) and among policymakers. A new label creates the illusion of progress. X-innovation also contributes to scholars' citation records. On the other hand, people contest the term (technological innovation) because of its hegemonic connotation. They coin alternative ones that often become a brand.

Notes

1. At about the same time in Britain, the Advisory Council for Applied Research and Development (ACARD) produced a short report (26 pages) titled *Industrial Innovation*, which espouses the discourse on technological innovation as demand–pull (rather than science–push).
2. A recent survey from the OECD reveals that one-third of countries (11 countries or 32 per cent) have a single ministry for both research and innovation (only 18 per cent had separate ministries) and 10 countries placed emphasis on support for business innovation concerning research policy (OECD, 2018a).
3. The first use of the concept "strategy" for innovation policy is Charles Carter and Bruce Williams (1959a).
4. "It is important that policy does not simply focus on supporting the creation of new knowledge" (OECD, 2009: 5).
5. Growth and competitiveness. In the 1990s, the OECD developed similar strategies (see Godin, 2004).
6. A question worth studying is to what extent the source of these typologies constitutes a borrowing to the

typology and dichotomy fundamental (basic) research *versus* applied research.

7. System Development Corporation (SDC) was a computer software company based in Santa Monica, California. Founded in 1955, SDC began as the systems engineering group for the SAGE air-defense system at the RAND Corporation. RAND spun off the group in 1957 as a non-profit organization. SDC became a for-profit corporation in 1969.

8. See http://www.oecd.org/gov/innovative-govern ment/a-framework-for-public-sector-innovation.htm.

9. See https://ec.europa.eu/growth/industry/innova tion/policy/public-sector_en.

10. In one place, the document talks of the "generality and ambiguity that attaches to this term [innovation] as well as to the hopes it arouses and the abuses to which it is often subject" (p. 11).

11. See https://ec.europa.eu/growth/industry/innova tion/policy/social_en.

12. See http://www.oecd.org/fr/cfe/leed/forum-social-innovations.htm.

13. Aant Elzinga and Andrew Jamison (1995) suggest such a change in policy priorities over time. Jean-Jacques Salomon contests the interpretation (1977: 60).

14. See https://ec.europa.eu/programmes/horizon2020/en/h2020-section/responsible-research-innovation.

Conclusion

During the last decades, innovation has become a highly valued concept. This book concentrates on the contribution of practitioners to the discourse. What about scholars? Scholars have no precursor role, but are members of a society that, after the Second World War, started to look at innovation with praise: practitioners, parliamentarians and congressmen, business organizations, the press – both popular and scientific – essayists, managers and engineers. The scholarly study of technological innovation began in the 1950s in several disciplines, at the same time as among practitioners: sociology,[1] geography,[2] anthropology,[3] history,[4] political science,[5] management[6] and economics[7] – education developed a bit later (in the 1960s).[8] This literature looks at research as a factor in innovation, diffusion of innovation, innovation and economic theory, innovation and policy. From the 1960s and 1970s onwards, a whole industry of conferences, books and articles developed and offered theories, frameworks and approaches (models) to make sense of technological innovation and contribute to public policies and firm strategies. Research centers were created and reviews began to appear, a sign of the field coming into existence. This literature did not constitute an organized field of study, not yet. But it set the tone for later theories.

It is within a policy context or background that thoughts on technological innovation developed. "Innovation connections" or networks were quite important, like that of the Department of Commerce,[9] the NSF[10] and the OECD.[11] It is in the footsteps of the US Department of Commerce that the Center for Policy Alternatives developed (Herbert Hollomon) and in the footsteps of the OECD that the SPRU was created (Chris Freeman). Many concepts used today come from the practitioners, such as the "innovation process" and "system of innovation", to name but a few.

Science policy has its historians (for example, Averch, 1985; Lucena, 2005; Salomon, 1977, 2000; Smith, 1990), but innovation policy has none. It is recognized today that the OECD has been responsible as a stimulator or catalyzer of thoughts on national science policy (Henriches and Laredo, 2013; Salomon, 1977: 43; 2000: 45, 51). Similarly, here I argue that innovation policy owes its existence to practitioners. Innovation theory does too. There is an assumption among scholars that political thinking is a second-order phenomenon, at best a matter of co-production between policymakers and scholars. This would make it an effect rather than a cause of the theoretical thinking. But the co-production itself is rarely studied, with a few exceptions (for example, Jamison, 1989; Mytelka and Smith, 1992; Sharif, 2006). Scholars' contributions dominate the narrative. I document here that the practitioners initiated a "national" discourse on technological innovation. The documents surveyed in this book are witness to the fact that policymakers and their consultants initiated and produced thoughts on technological innovation that, to varying degrees, preceded and influenced scholars' representations of innovation in the following decades. Public organizations were among the first to produce titles on technological innovation, providing sources that scholars could draw from. Subsequent scholars' theories served to explain and articulate the then existing views (for example, Freeman, 1974).

The history of technological innovation as an idea is generally told from an academic point of view, focusing on the emergence and development of theories of innovation. I am guilty of writing such a history as well. The scholars' history is also written, most of the time, from a research perspective. Innovation is conflated with science and research (hence the acronym STI). This is a construct that many do not share.

This book suggests that science and scientific research as factors of economic growth gave rise to the idea of technological innovation. But in turn, technological innovation led to the marginalization of research. The discourse on innovation served the discourse on science for a while and vice versa. But the success of the former over that of science is inclusiveness. Innovation includes scientists, engineers, designers, entrepreneurs and firms. It also includes customers (needs, users). It is also a matter of diverse policies. This does not mean that research has disappeared from public discourse. To some extent, research remains a key factor in technological innovation, according to some of the discourses. The original meaning of the term "technological innovation" may explain this conception: "science applied" to industry, research as a source of progress, and the cultural value of science to society, namely the place of "science-based industries" in the hierarchy of technological innovation. Yet, the conception competes, for better or worse, with that of the market: that technological innovation is the commercialization of new goods, and the principal agents in the process are firms, not scientists.

Notes

1. Eugene Wilkening, George Beal and Joe Bohlen, Everett Rogers, James Coleman.
2. Torsten Hagerstrand, Lawrence Brown.
3. Mary Hodgen, Homer Barnett.
4. Abbot Usher, Elting Morison, Rupert Maclaurin.
5. Karl Deutsch.
6. Arthur Cole's Research Center in Entrepreneurial History, Peter Drucker, Herbert Simon, Selwyn Becker.
7. Moses Abramovitz, Walter Rostow, François Perroux, Simon Kuznets, William Fellner, Jacob Schmookler, Zvi Griliches, Vernon Ruttan, Frederik Scherer, Marian Bowley and Audrey Donnithorne (work conducted between 1953 and 1960 at University College, London), Jean-Louis Maunoury.
8. Matthew Miles, Richard Carlson.
9. Herbert Hollomon, Robert Charpie, Daniel de Simone,

Aaron Gellman and Arthur D. Little Inc. (Donald Schon, Michael Michaelis).

10. Consultants like Edwin Mansfield and the National Planning Association.
11. Chris Freeman, Keith Pavitt and experts from Member countries.

Appendices

Appendix 1

Inclusive policy at the OECD

Science policy

The document produced for the first OECD ministerial conference on science in 1963 stated: "the relationship between a national policy for economic development and a national policy for scientific research and development is one of the essential subjects for study" (OECD, 1963b: 52). To the OECD, what is needed is a dialogue between those responsible for economic policy and those responsible for science policy (pp. 69–73). Science policy is not integrated. As another OECD document of the same year put it: "there is a great need for studies of the several fields and ways in which science and policy interact, and there is a need above all for a continuing and intimate working relationship between officials responsible for science policy and other policy makers" (OECD, 1963a: 26–7). "National policies in other fields must take account of the achievements and expectations of science and technology": economic policy, social policy, military policy, foreign policy, aid policy (p. 26).

One of the OECD studies that most explicitly carried a system approach was *The Research System*, published in three volumes between 1972 and 1974 under the direction of Jean-Jacques Salomon.[1] The study framed the central issue of the system approach in terms of a dichotomy between two periods, as had a previous report (the Piganiol Report: OECD, 1963a: 18): the policy for science period as the expansion of research per se, versus the science for policy period where "developing national research potential [is] generally regarded as synonymous with national innovation potential" (OECD, 1974b: 168). The study looked at elements of the research system in ten countries, large and small: organization, financing, application of science (or innovation), government research, university–industry relations and international dimensions. Because research is not an autonomous system, according to the authors, the document "put emphasis on the institutional context in which research is conducted. One of the most delicate problems of science policy is how to influence the process by which scientific discoveries are transformed into useful applications and how to contribute, in some way or another, towards bringing the supply of science into closer harmony with the demand of society" (OECD, 1971: 16). "The whole problem of university research consists in the break-up of its institutional framework" (pp. 17–18, 20):

> The needs of fundamental research depend primarily on the talent available and the fields opened up by the

unsolved (or unformulated) problems of science itself. The needs of applied research and development, on the other hand, depend primarily on the problems which the industrial system sets itself. There is no hermetic seal between the first type of problem and the second, the terms of each being renewed or changed by the progress made by the other on the basis of a certain degree of osmosis between the university and industry and that is precisely why it is better to speak of a "research system" rather than a juxtaposition or hierarchy of different forms of research.

As a major conclusion from the study, *The Research System* suggested: "scientific and technological research, viewed from an institutional approach, cannot be separated from its political, economic, social and cultural context" (p. 22).

Science and technology policy

At the request of the OECD Secretary-General, a Group on New Concepts of Science Policy made a reassessment of science policy between 1969 and 1971. The Group used "the expression 'science policy' for reasons of brevity, it should be taken throughout to mean policies for the natural sciences, the social sciences, and technology" (OECD, 1971: 17).

The group identified three phases of science policy (pp. 39–40): (1) Second World War–1960: faith in science (with the "magic 3 per cent research ratio" as a kind of "touchstone of scientific success", (pp. 40, 45); "science policy in most countries comes down to research policy":

"financing research regardless of the aims it may serve", p. 45); (2) 1961–7: economic growth; (3) 1960s: disenchantment (environmental and social deterioration resulting from technology).

We need to establish, claimed the report, "an assessment process embracing the *whole* economy and relating the *whole* of technology to the *whole* social and natural environment" (p. 79):

> The new tasks faced by science and technology are more complex and multivariant than the old ones, involving economic, social, cultural and psychological aspects, as well as more strictly technical ones. (p.92)

> We can no longer consider technological progress as an independent variable. Science and technology are an integral part of social and economic development, and we believe that this implies a much closer relationship between policies for science and technology and all socio-economic concerns and governmental responsibilities than has existed in the past. (p. 96)

To the group "the composition of output is likely to shift from the production of private goods for the market to the production of 'public goods'" (pp. 21–2). "By its very success the economic approach has proved its limitations" (p. 26).[2] There is "need to approach the question of development of societies more comprehensively, going beyond exclusively economic considerations" (p. 31), namely in terms of "interactions between technical, economic, psychological, and sociological factors" (p. 34). As Harvey Brooks put it in the cover letter accompanying the report, "the concept of science policy as a discrete and isolable element of government policy will tend to be replaced by a much broader view, one in which there is a close feedback relationship between technological opportunities and social goals". Science policy is too fragmented and uncoordinated. "A more comprehensive approach", namely "science policy as an integral factor in overall public policy" is needed (p.12). "Purely economic solutions are insufficient . . . Science policy must be much more broadly conceived than in the past" (pp. 30, 36):

> First, the different elements of science policies were usually treated independently of each other; second, science policies themselves were often treated in relative isolation from other policy decisions. (p. 47)

> [Now], science and technology are an integral part of social and economic development, and we believe that this implies a much closer relationship between policies for science and technology and all socio-economic concerns and governmental responsibilities than has existed in the past. (p. 96)

This was also the message of the third (1968) and fourth (1971) OECD ministerial meetings. It is "now accepted that technological progress is an essential component of economic growth". Yet "the system of industrial innovation has its limitations" (OECD, 1970: 2). Demand or market goods do not respond sufficiently to social requirements (public goods) and have undesirable repercussions. This requires more technology ("a more intensive scientific and technological effort", but through "a healthy system of industrial innovation . . . responsive to the social needs" (p. 3). Such a system is composed of industrial research and development, entrepreneurial management, fundamental research (universities), financial resources, mobility of scientists and engineers, flexibility of industrial structure, climate for competition, international market integration, and technological specialization ("no one company or country can hope to exploit all of the opportunities", p. 4). "The task of the 1970s will be to translate new policy objectives into demands for innovations designed to solve new problems and meet new needs, and also by introducing standards and penalties for socially undesirable and damaging activities" (p. 4). This "involves decisions on *total* [my italics], national resources allocation which concern social and economic policy" (p. 5). "In the future . . . the definition of science policy objectives must result from the interaction between scientific and technological opportunities, and the *totality* [my italics] of economic and social needs" (p. 6).

But in general, the statement that science and technology policy is not integrated means that it is not integrated into economic policy. Over the years, the main policy coordination expected from science policy was with economics. Because "science and technology policies have usually been defined and implemented independently of economic policies", a report from 1980 titled

Technical Change and Economic Policy recommended that science and technology policies be better integrated with economic policies (OECD, 1980: 12, 93):

If there is little justification for assuming limits to science and technology, there are limitations imposed by political, economic, social and moral factors which may retard, inhibit or paralyze both scientific discovery and technical innovation. The most intractable problems lie not in the potential of science and technology as such, but rather in the capacity of our economic systems to make satisfactory use of this potential.

The committee responsible for the report recommended a "better integration of the scientific and technical aspects of public policy, and the social and economic aspects" and "much closer links regarding such government functions as providing for national defence, agricultural productivity, health, energy supply, and protecting the environment and human safety". To the committee, "the organizations that propose and carry out science and technology policies tend to stand separate from offices at a comparable level concerned with the more legal and economic aspects of policy" (p. 96).

Notes

1. An early OECD document on system analysis *à la* Forrester and its application to scientific and technical system is OECD (1974a). However, this document has never been published. Another OECD study using the idea of system is OECD (1972b).

2.   "The accuracy and sensitivity of the market mechanism as an indicator of the aspirations of society can be called into question . . . A range of new needs are not readily expressible as market demand" (p. 27).

Appendix 2

OECD policy instruments

Direct financing of enterprises' R&D projects
  Grants
  Loans
  Joint-venture risk-sharing
  Equity sharing
  Setting up new enterprises
Aid to inventions
  General fiscal incentives
  Incentives for extra-curricular R&D
  Legal aid for employees
  Technical assistance
  Licensing aid and amendments to patent
    regulations
  Financial aid for development
  Financial aid for setting up new
    enterprises
Major (or government) programs
  Mobilizing or setting up public R&D
    capabilities
  Mobilizing private R&D capabilities
  Public markets
  Combining measures of industrial policy
    Grants for equipment
    New enterprises
    Restructuring industries
  Directing the efforts of central scientific and
    technological policy institutions
Acting on market forces

Coordination and aggregation of demand; financing feasibility

Consumer assistance for quality control operations

Regulations and standards

Performance standards, standard improvements, consumer protection and regulations

Public purchasing

Industrial policy

Updating and rationalizing equipment

Introduction of new techniques in production structures

Productivity studies aiming at innovation

Establishment of new industries

Small business adaptation

Initiatives related to government programs

Public R&D access to industry

Long-term research for industry's needs

Establishment of physical and industrial standards

New technology feasibility studies

Provisions of clauses establishing:

Cooperation with industry for the management of R&D programs

Systematic analysis procedures for evaluating needs

Contract research procedures

Transfer of technology and know-how from government R&D

Transfer institutions linked with government programs

Dissemination and technology transfer services
for big science institutions
Patent regulations
Patent offices linked with public institutions
Central brokerage institutions
Scientific and technology transfer between
industries
Central brokerage institutions
Institutions for disseminating and promoting
technical achievements for the benefit of
SMEs
Research associations

Appendix 3

SPRU policy instruments

Direct
Funding industrial R&D
Procurement
Tax credits

Indirect
Contributions to international programs
Funding public technical institutes
Information services
Industrial standards
Patents

Other
Regional policy
Sectoral policy
Structural policy (employment)
Taxation
Public R&D

Appendix 4

CPA policy instruments

Initiating mechanisms (innovation process)
Stimulating innovation by working through market forces
Reducing the costs to firms of undertaking innovative activities
Reducing the probability of technical or commercial failure
Increasing the rewards to the firm for successful innovation
Encouraging innovation via market invasion
Restructuring an industrial sector
Influencing the organization and management of individual forms

Sustaining mechanisms (resources)
Influencing the availability, utilization and mobility of managerial and technical manpower
Assisting institutions (universities, research institutes, private consulting firms, industries and governments) with regard to the generation and utilization of technical knowledge
Increasing and transmission and transfer of technical knowledge between institutions

Restructuring mechanisms (consequences)
Ameliorating the adverse consequences of technology with respect to the environment and natural resources
Influencing labor's receptivity to technological

change and internalizing the human costs associated with innovation activity

Appendix 5

CRS policy instruments

Taxation
Venture capital
Foreign trade
Federal R&D
Economic regulation
Environment, health, safety and consumer
  regulation
Antitrust policy
Patent policy
Labor and manpower
Federal procurement
Information dissemination
State/local government innovation
Corporate organization for innovation
General

References

Abramovitz, Moses (1952), Economics of Growth, in Bernard F. Haley (ed.), *A Survey of Contemporary Economics*, vol. 3, Homewood (Ill.): Richard D. Irwin: 152–78.

Alford, Leon P. (1929), *Technical Changes in Manufacturing Industries, in National Bureau of Economic Research, Recent Economic Changes in the United States*, New York: McGraw-Hill: 96–166.

Allen, Thomas J., James M. Utterback, Marvin A. Sirbu, Nicholas A. Ashford and J. Herbert Hollomon (1978), Government Influence on the Process of Innovation in Europe and Japan, *Research Policy*, 7: 124–49.

Argyris, Chris (1965), *Organization and Innovation*, Homewood (Ill.): Richard D. Irwin.

Arndt, Heinz Wolfgang (1978), *The Rise and Fall of Economic Growth*, Chicago (Ill.): University of Chicago Press.

Arndt, Heinz Wolfgang (1981), Economic Development: A Semantic History, *Economic Development and Cultural Change*, 29 (3): 457–66.

Arndt, Heinz Wolfgang (1987), *Economic Development: The History of an Idea*, Chicago (Ill.): University of Chicago Press.

Arthur D. Little Inc. (1963), *Patterns and Problems*

of Technical Innovation in American Industry, Report to the National Science Foundation, Washington: National Science Foundation.

Arthur D. Little Inc. (1965), *Management Factors Affecting Research and Exploratory Development*, Report to the US Department of Defense, Cambridge (Mass.): Arthur D. Little Inc.

Arthur D. Little Inc. (1968), *A Model for Innovation Adoption in Public School Districts*, Report to the Office of Education, US Department of Health, Education and Welfare, Cambridge (Mass.): Arthur D. Little Inc.

Arthur D. Little Inc. and Industrial Research Institute (1973), *Barriers to Innovation in Industry: Opportunities for Public Policy Changes*, Report to the National Science Foundation, Washington: National Science Foundation.

Averch, Harvey A. (1985), *A Strategic Analysis of Science & Technology Policy*, Baltimore (Md.): Johns Hopkins University Press.

Battelle-Columbus Laboratories (1973), *Interactions of Science and Technology in the Innovative Process: Some Case Studies*, Report to the National Science Foundation, Washington: National Science Foundation.

Becker, Selwyn W. (1964), *The Innovative Organization*, Selected Paper no. 14, Graduate School of Business, University of Chicago.

Becker, Selwyn W. and Thomas L. Whisler (1967), The Innovative Organization: A Selective View of Current Theory and Research, *Journal of Business*, 40 (4): 462–9.

Bichowsky, F. Russell (1942), *Industrial Research*, New York: Chemical Publishing.

Blackett, Patrick M.S. (1968), Memorandum to the Select Committee on Science and Technology, *Nature*, 219, September 14: 1107–10.

Borras, Susana and Charles Edquist (2013), The Choice of Innovation Policy Instruments, *Technological Forecasting and Social Change*, 80: 1513–22.

Boulding, Kenneth E. (1953), *The Organizational Revolution*, New York: Harper.

Boulding, K.E. (1946), The Principles of Economic Progress, in *The Economics of Peace*, New York: Prentice-Hall: 73–101.

Branscomb, Lewis M. and James H. Keller (eds) (1998), *Investing in Innovation: Creating a Research and Innovation Policy that Works*, Cambridge (Mass.): MIT Press.

Bright, James R. (1964), *Research, Development, and Technological Innovation: An Introduction*, Homewood (Ill.): Richard D. Irwin.

British Advisory Council for Applied Research and Development (1979), *Industrial Innovation*, by a committee chaired by A. Sprinks (ICI), London: HSMO.

British Advisory Council on Scientific Policy (1964), *Annual Report 1963–1964*, London: HMSO.

British Central Advisory Council for Science and Technology (1968), *Technological Innovation in Britain*, London: HMSO.

British Department of Scientific and Industrial

Research (1927), *Factors in Industrial and Commercial Efficiency, Committee on Industry and Trade*, Part I, Chapter 4, London: HMSO.

Brozen, Yale (1942), *Some Economic Aspects of Technological Change*, Thesis, Department of Economics, University of Chicago.

Burns, Robert O. (1975), *Innovation: The Management Connection*, Lexington (Mass.): Lexington Books.

Burns, Tom (1956), The Social Character of Technology, *Impact of Science on Society*, 7 (3): 147–65.

Burns, Tom and George M. Stalker (1955), The Management of Innovation, *Research Applied in Industry*, 8 (7): 247–51.

Burns, Tom and George M. Stalker (1961 [1996]), *The Management of Innovation*, London: Tavistock Publications.

Bush, Vannevar (1945 [1995]), *Science: The Endless Frontier*, North Stratford: Ayer Co..

Caffrey, John (1965), *The Innovation Matrix*, Santa Monica (Calif.): System Development Corporation.

Carson, Carl S. (1975), The History of the United States National Income and Product Accounts: The Development of an Analytical Tool, *Review of Income and Wealth*, 1: 153–81.

Carter, Charles F. and Bruce R. Williams (1957), *Industry and Technical Progress: Factors Governing the Speed of Application of Science*, London: Oxford University Press.

Carter, Charles F. and Bruce R. Williams (1958),

Investment in Innovation, London: Oxford University Press.

Carter, Charles F. and Bruce R. Williams (1959a), *Science in Industry: Policy for Progress*, London: Oxford University Press.

Carter, Charles F. and Bruce R. Williams (1959b), The Characteristics of Technically Progressive Firms, *Journal of Industrial Economics*, 7 (2): 87–104.

Center for Policy Alternatives (1975), *National Support for Science & Technology: An Examination of Foreign Experience*, Cambridge (Mass.): MIT.

Center for Policy Alternatives (1978), *Government Involvement in the Innovation Process*, Washington: Office of Technology Assessment.

Charpie, Robert A. (1967a), Technological Innovation and Economic Growth, in National Academy of Sciences, *Applied Science and Technological Progress*, Washington: 357–64.

Charpie, Robert A. (1967b), The Business End of Technology Transfer, in National Science Foundation, *Proceedings of a Conference on Technology Transfer and Innovation*, 15–17 May 1966, NSF 67-5, Washington: NSF: 46–52.

Charpie, Robert A. (1970), Technological Innovation and the International Economy, in M. Goldsmith (ed.), *Technological Innovation and the Economy*, London/New York: Wiley: 1–10.

Colm, Gerhard (1958), Discussion, in National Science Foundation, *Proceedings of a Conference on Research and Development and its Impact on the Economy*, NSF 58-36, Washington: NSF: 149–52.

Coyle, Diane (2014), *GDP: A Brief but Affectionate History*, Princeton (N.J.): Princeton University Press.

Croome, Honor (1960a), Human Problems in Industry, *Nature*, 4723, 7 May: 419–20.

Croome, Honor (1960b), *Human Problems of Innovation*, London: HMSO.

Denison, Edward F. (1962), *The Sources of Economic Growth in the United States and the Alternatives Before Us*, New York: Committee for Economic Development.

Dixon, Roland B. (1928), *The Building of Cultures*, New York: Charles Scribner.

Douglas, Paul H. (1930), Technological Unemployment, *Bulletin of the Taylor Society*, 15 (6): 254–70.

Duckworth, John Clifford (1965a), Incentives to Innovation and Invention, *Electronics and Power*, 11 (6): 186–90.

Duckworth, John Clifford (1965b) The Process of Technological Innovation, *Proceedings of the Institutions of Electronic and Radio Engineers*, 3 (3): 89–94.

Duckworth, John Clifford (1970), The Role of Government, in Maurice M. Goldsmith (ed.), *Technological Innovation and the Economy*, London/New York: Wiley: 111–17.

Edgerton, David (2004), The Linear Model did not Exist, in Karl Grandin, Nina Worms and Sven Widmalm (eds), *The Science–Industry Nexus: History, Policy, Implications*, Sagamore Beach (Mass.): Science History Publications: 31–57.

Elzinga, Aant and Andrew Jamison (1995), Changing Policy Agendas in Science and Technology, in Sheila Jasanoff et al. (eds), *Handbook of Science and Technology Studies*, California: Sage: 572–97.

European Commission (2010), Communication from the Commission to the European Parliament, the Council, the European Economic and Social Committee and the Committee of the Regions, Europe 2020 Flagship Initiative Innovation Union, COM(2010) 546, Brussels.

European Parliament (2016), *Innovation Policy: Building the EU Innovation Policy Mix*, European Parliamentary Research Service, European Union.

Ewell, Raymond H. (1955), Role of Research in Economic Growth, *Chemical and Engineering News*, 33 (29): 2980–85.

Fabricant, Solomon (1965), *Measurement of Technological Change*, Seminar on Manpower Policy and Program, US Department of Labor, July 1965.

Fabricant, Solomon, Michael Schiff, Joseph G. San Miguel and Shahid L. Ansari (1975), *Accounting by Business Firms for Investment in Research and Development, Section III: The Literature on Accounting for Innovation*, Report to the NSF: New York: New York University.

Fagerberg, Ian, Ben R. Martin and Esben Sloth Andersen (eds) (2013), *Innovation Studies: Evolution & Future Challenges*, Oxford: Oxford University Press.

Flanagan, Kiron, Elvira Uyarra and Manuel Laranja (2011), Reconceptualizing the "Policy Mix" for Innovation, *Research Policy*, 40: 702–13.

Freeman, Chris (1963), The Plastics Industry: A Comparative Study of Research and Innovation, *National Institute Economic Review*, 26: 22–49.

Freeman, Chris (1974), *The Economics of Industrial Innovation*, Middlesex: Penguin Books.

Furnas, Clifford C. (ed.) (1948), *Research in Industry: Its Organization and Management*, Princeton (N.J.): D. Van Nostrand.

Gaglio, Gerald (2012), Du lien entre l'étude sociologique de l'innovation et la sociologie: une lecture simmelienne, *Cahiers de recherche sociologique*, 54: 49–72.

Gardner, John W. (1960), National Goals in Education, in President's Commission on National Goals (1960), *Goals for Americans*, New York: Prentice-Hall: 81–100.

Gardner, John W. (1963), *Self-Renewal: The Individual and the Innovative Society*, New York: Norton.

Gellman, Aaron (1966), A Model of the Innovative Process (as Viewed from a Non-Science Based Fragmented Industry), in National Science Foundation, *Technology Transfer and Innovation*, Proceedings of a Conference Organized by the National Planning Association and the NSF in May, 15–17, NSF 67-5, Washington: NSF: 11–20.

Gellman, Aaron J. (1970), Market Analysis and Marketing, in Maurice M. Goldsmith (ed.),

Technological Innovation and the Economy, London/New York: Wiley: 129–37.

Gellman, Research Associates (1974), *Economic Regulation and Technological Innovation: A Cross National Literature Survey and Analysis*, NSF Rep. PB-233085/AS.

Gellman Research Associates (1976), *Indicators of International Trends in Technological Innovation*, Report to the NSF, Washington: NSF.

Gilfillan, S. Colum (1935), *The Sociology of Invention*, Cambridge (Mass.): MIT Press.

Godin, Benoît (2002), Technological Gaps: An Important Episode in the Construction of Science and Technology Statistics, *Technology in Society*, 24: 387–413.

Godin Benoît (2004), The New Economy: What the Concept Owes to the OECD, *Research Policy*, 33: 679–90.

Godin, Benoît (2005), *Measurement and Statistics on Science and Technology: 1920 to the Present*, London: Routledge.

Godin, Benoît (2009), *National Innovation System (II): Industrialists and the Origins of an Idea*, Project on the Intellectual History of Innovation, Montreal: INRS. Available at http://www.csiic.ca/PDF/IntellectualNo4.pdf.

Godin, Benoît (2015), *Innovation Contested: The Idea of Innovation Over the Centuries*, London: Routledge.

Godin, Benoît (2017), *Models of Innovation: The History of an Idea*, Boston (Mass.): MIT Press.

Godin, Benoît (2019), *The Invention of Technological*

Innovation: Innovation Language, Discourse and Ideology in Historical Perspective, Cheltenham: Edward Elgar Publishing.

Godin, Benoît and Joseph Lane (2013), Pushes and Pulls: The Hi(S)tory of the Demand-Pull Model of Innovation, *Science, Technology and Human Values*, 38 (5): 621–54.

Godin, Benoît and Désirée Schauz (2016), The Changing Identity of Research: A Cultural and Conceptual History, *History of Science*, 54 (3): 276–306.

Goldsmith, Maurice M. (ed.) (1970), *Technological Innovation and the Economy*, London/New York: Wiley: xiii–xvii.

Gourvitch, Alexander (1940), *Survey of Economic Theory on Technological Change and Employment*, National Research Project on Reemployment Opportunities and Recent Changes in Industrial Techniques, Philadelphia (Pa.): Works Progress Administration.

Griliches, Zvi (1958), Research Costs and Social Return: Hybrid Corn and Related Innovation, *Journal of Political Economy*, 66 (5): 419–31.

Gruber, William H. and Donald G. Marquis (eds) (1969), *Factors in the Transfer of Technology*, Cambridge (Mass.): MIT Press.

Hage, Jerald and Michael Aiken (1970), *Social Change in Complex Organizations*, New York: Random House.

Hansen, Alvin H. (1921), The Technological Interpretation of History, *Quarterly Journal of Economics*, 36 (1): 72–83.

Hansen, Alvin H. (1932), The Theory of Technological Progress and the Dislocation of Employment, *American Economic Review*, 22 (1): 25–31.

Hansen, Alvin H. (1939), Economic Progress and Declining Population Growth, *American Economic Review*, 29 (1): 1–15.

Hansen, John A., James I. Stein and Thomas S. Moore (1984), *Industrial Innovation in the United States: A Survey of Six Hundred Companies*, Center for Technology and Policy, Report to the NSF, Boston (Mass): Boston University.

Harbridge House Inc. (1975), *Legal Incentives to Private Investment in Technological Innovation*, Office of Experimental R&D Incentives, NSF.

Havelock, Ronald G. (1969), *Planning for Innovation Through Dissemination and Utilization of Knowledge*, Center for Research on Utilization of Scientific Knowledge (CRUSK), Ann Arbor (Mich.): Institute for Social Research, University of Michigan.

Havelock, Ronald G. (1970), *A Guide to Innovation in Education*, Center for Research on Utilization of Scientific Knowledge (CRUSK), Ann Arbor (Mich.): Institute for Social Research, University of Michigan.

Havelock, Ronald G. and Mary C. Havelock (1973), *Educational Innovation in the United States*, report to the National Institute of Education, US Office of Education, Washington.

Henderson, James B. (1927), Invention as a Link

in Scientific and Economic Progress, *Nature*, 120 (3024), 15 October: 550–53, 588–91.

Henriches, Luisa and Philippe Laredo (2013), Policy-Making in Science Policy: The "OECD Model" Unveiled, *Research Policy*, 42: 801–16.

Hertz, David Bendel (1965), The Management of Innovation, *Management Review*, April: 49–52.

Hicks, John R. (1966), Growth and Anti-Growth, *Oxford Economic Papers*, 18 (3): 257–69.

Hildred, William H. and Leroy A. Bengston (1974), *Surveying Investment in Innovation*, University of Denver, Colorado, Report to the NSF.

Hill, Christopher T., John A. Hansen and James H. Maxwell (1982), *Assessing the Feasibility of New Science and Technology Indicators*, Center for Policy Alternatives, Report to the NSF, MIT, Cambridge (Mass.): MIT.

Hill, Christopher T., John A. Hansen and James I. Stein (1983), *New Indicators for Industrial Innovation*, Center for Policy Alternatives, Report to the NSF, Cambridge (Mass.): MIT.

Hill, Christopher T. and James M. Utterback (eds) (1979), *Technological Innovation for a Dynamic Economy*, Center for Policy Alternatives, New York: Pergamon Press.

Holland, Maurice (1928), Research, Science and Invention, in F.W. Wile (ed.), *A Century of Industrial Progress*, American Institute of the City of New York, New York: Doubleday, Doran and Co.: 312–34.

Hollomon, John Herbert (1960 [1991]),

Engineering's Great Challenge – The 1960s, in Hedy E. Sladovich (ed.), *Engineering as a Social Enterprise*, Washington: National Academy Press: 104–10.

Hollomon, John Herbert (1965a [1968]), Science and Innovation, in Richard A. Tybout (ed.), *Economics of Research and Development*, Columbus (Ohio): Ohio State University Press: 251–7.

Hollomon, John Herbert (1965b), Science and the Civilian Technology, in Aaron W. Warner, Dean Morse and Alfred S. Eichner (eds), *The Impact of Science on Technology*, New York: Columbia University Press: 118–42.

Hollomon, John Herbert (1967), Technology Transfer, in National Science Foundation, *Technology Transfer and Innovation*, Proceedings of a Conference Organized by the National Planning Association and the National Science Foundation, May, 15–17, 1966, NSF 67-5, Washington: NSF: 32–6.

Hollomon, John Herbert (1968), Creative Engineering and the Needs of Society, in Daniel V. Simone (ed.), *Education for Innovation*, Oxford: Pergamon Press: 23–30.

Hollomon, John Herbert (1979), Government and the Innovation Process, *Technology Review*, May: 30–37.

Hollomon, John Herbert (1980), Appropriate Role of Government in Innovation, in W. Novis Smith (ed.), *Innovation and US Research: Problems and Recommendations*, Washington: American Chemical Society: 197–204.

Hounshell, David (2000), The Medium is the Message, or How Context Matters: The RAND Corporation Builds an Economics of Innovation, 1946–62, in Agatha C. Hughes and Thomas P. Hughes (eds) (2000), *Systems, Experts and Computers: The System Approach in Management and Engineering, World War II and After*, Cambridge (Mass.): MIT Press: 255–310.

Illinois Institute of Technology (1968), *Technology in Retrospect and Critical Events in Science (TRACES)*, Washington: National Science Foundation.

Jamison, Andrew (1989), Technology's Theorists: Conceptions of Innovation in Relation to Science and Technology Policy, *Technology and Culture*, 30 (3): 505–33.

Jerome, Harry (1934), *Mechanization in Industry*, New York: National Bureau of Economic Research.

Jewkes, John (1960), How Much Science, *The Economic Journal*, 70 (277): 1–16.

Jewkes, John, David Sawers and Richard Stillerman (1958), *The Sources of Invention*, London: Macmillan.

Johnson, Ann (2008), What if We Wrote the History of Science from the Perspective of Applied Science?, *Historical Studies in the Natural Sciences*, 38 (4): 610–20.

Kennedy, Charles and Anthony P. Thirlwall (1972), Surveys in Applied Economics: Technical Progress, *Economic Journal*, 82 (325): 11–72.

Kreilkamp, Karl (1971), Hindsight and the Real

World of Science Policy, *Science Studies*, 1 (1): 43–66.

Kuznets, Simon (1929), Retardation of Industrial Growth, *Journal of Business History*, 2: 534–60.

Kuznets, Simon (1930 [1967]), Retardation of Industrial Growth, in Simon Kuznets (ed.), *Secular Movements in Production and Prices: Their Nature and Their Bearing Upon Cyclical Fluctuations*, Boston (Mass.): Houghton Mills: 1–58 [New York: Augustus M. Kelley].

Kuznets, Simon (1959), The Problem of Measurement, in Bureau international de recherche sur les implications sociales du progrès technique, *Social, Economic and Technological Change: A Theoretical Approach*, Paris: Presses universitaires de France: 153–92.

Kuznets, Simon (1966), *Modern Economic Growth: Rate, Structure, and Spread*, New Haven (Conn.): Yale University Press.

Langrish, John, Michael Gibbons, William G. Evans and Frederic Raphael Jevons (1972), *Wealth from Knowledge: Studies of Innovation in Industry*, London: Macmillan.

Layton, Christopher (1972), *Ten Innovations: An International Study on Technological Development and the Use of Qualified Scientists and Engineers in Ten Industries*, London: Allen & Unwin.

Lederer, Emil (1938), *Technical Progress and Unemployment: An Inquiry into the Obstacles to Economic Expansion*, International Labour Office, Geneva: P.S. King.

Locke, H. Brian (1976), Innovation by Design, *Electronics and Power*, 22 (9): 611–14.

Lonigan, Edna (1939), The Effect of Modern Technological Conditions upon the Employment of Labor, *American Economic Review*, 29 (2): 246–59.

Lorsch, Jay W. and Paul R. Lawrence (1965), Organizing for Product Innovation, *Harvard Business Review*, 43: 109–22.

Lucena, Juan C. (2005), *Defending the Nation: US Policymaking to Create Scientists and Engineers fom Sputnik to the War Against Terrorism*, Lanham (Mass.): University Press.

Maclaurin, W. Rupert (1949), *Invention and Innovation in the Radio Industry*, New York: Macmillan.

Maclaurin, W. Rupert (1953), The Sequence from Invention to Innovation and its Relation to Economic Growth, *Quarterly Journal of Economics*, 67 (1): 97–111.

Manchester Joint Research Council (1954), *Industry and Science*, Manchester: Manchester University Press.

Mansfield, Edwin (1961), Diffusion of Technological Change, *Review of Data on Research & Development*, 31, NSF 61–52, October.

Mansfield, Edwin (1962), Entry, Gibrat's Law, Innovation, and the Growth of Firms, *American Economic Review*, 52 (5): 1023–51. No explicit definition: "First to introduce the significant new processes and products" (p.1035).

Mansfield, Edwin (1963), Innovation in Individual

Firm, *Review of Data on Research & Development*, 34, NSF, June: 62–16.

Mansfield, Edwin (1972), Contribution of R&D to Economic Growth in the United States, *Science*, 175 (4021): 477–86.

Mansfield, Edwin and John Rapoport (1975), The Costs of Industrial Product Innovations, *Management Science*, 21 (12): 1380–86.

Mansfield, Edwin, John Rapoport, Jerome Schnee, Samuel Wagner and Michael Hamburger (1971), *Research and Innovation in the Modern Corporation*, New York: Norton.

Mansfield, Edwin, John Rapoport, Anthony Romeo, Samuel Wagner and George Beardsley (1977), Social and Private Rates of Return from Industrial Innovations, *The Quarterly Journal of Economics*, 91 (2): 221–40.

Masoon, Ehsan (2016), *The Great Invention: The Story of the GDP and the Making and Unmaking of the Modern World*, New York: Pegasus Books.

Massell, Benton F. (1960), Capital Formation and Technological Change in United States Manufacturing, *Review of Economics and Statistics*, 42 (2): 182–88.

Mechling, Kenneth R. (1969), *A Strategy for Stimulating the Adoption and Diffusion of Science Curriculum Innovations among Elementary School Teachers*, Office of Education, US Department of Health, Education and Welfare, Clarion (Pa.): Clarion State College.

Mees, C.E. Kenneth (1920), *The Organization*

of *Industrial Scientific Research*, New York: McGraw-Hill.

Michaelis, Michael (1964), Obstacles to Innovation, *International Science and Technology*, November: 40–46.

Michaelis, Michael (1967), The Environment for Innovation, in US National Science Foundation, *Technology Transfer and Innovation*, Proceedings of a Conference Organized by the National Planning Association and the National Science Foundation, 15–17 May 1966, NSF 67-5, Washington: National Science Foundation: 76–81.

Minasian, Jora R. (1969), R&D, Production Functions and Rates of Return, *American Economic Review*, 59 (2): 80–85.

Mogee, Mary and Wendy Schacht (1980), *Industrial Innovation*, Congressional Research Service.

Morton, Jack A. (1971), *Organizing for Innovation: A Systems Approach to Technical Management*, New York: McGraw Hill.

Mottur, Ellis Robert (1968), *The Processes of Technological Innovation: Conceptual Systems Model*, report prepared for the Office of Invention and Innovation and the Arms Control and Disarmament Agency (Department of State), Washington: George Washington University's Program of Policy Studies in Science and Technology.

Mowery, David and Nathan Rosenberg (1979), The Influence of Market Demand Upon Innovation: A Critical Review of Some Recent Empirical Studies, *Research Policy* 8: 102–53.

Mueller, Robert K. (1971), *The Innovation Ethic*, New York: American Management Association.

Mueller, Willard F. (1957), A Case Study of Product Discovery and Innovation Costs, *Southern Economic Journal*, 24 (1): 80–86.

Myers, Sumner (1965), Attitude and Innovation, *International Science and Technology*, 46: 91–103.

Myers, Sumner (1967), *Industrial Innovations: Their Characteristics and Their Scientific and Technical Information Bases*, Washington: National Planning Association.

Myers, Sumner and Donald G. Marquis (1969), *Successful Industrial Innovations: A Study of Factors Underlying Innovation in Selected Firms*, NSF 69-17, Washington: National Science Foundation.

Myers, Sumner and Eldon E. Sweezy (1965), *Federal Incentives for Innovation: Why Innovations Falter and Fall – A Survey of 200 Cases*, Report to the NSF, Denver (Colo.): Denver Research Institute.

Mytelca, Lynn K. and Keith Smith (1992), Policy Learning and Innovation Theory: An Interactive and Co-Evolving Process, *Research Policy*, 31: 1467–79.

National Bureau of Economic Research (NBER) (1962), *The Rate and Direction of Inventive Activity: Economic and Social Factors*, Princeton (N.J.): Princeton University Press.

Nature (1979), Innovation: What's in a Word?, 282, 8, November: 119.

Nelson, Richard R. (1959a), The Economics of Invention: A Survey of the Literature, *Journal of Business*, 32 (2): 101–27.

Nelson, Richard R. (1959b), The Simple Economics of Basic Research, *Journal of Political Economy*, 67: 297–306.

Noll, Roger G. (1975), *Government Policies and Technological Innovation*, 5 volumes, Pasadena, (Calif.): Final Report, California Institute of Technology.

Novick, David (1960), What Do We Mean by R&D?, *California Management Review*, 2 (3) 9–24.

Novick, David (1965), The ABC of R&D, Challenge, 13 (5): 9–13.

OECD (1958), *Standardized System of National Accounts*, Paris: OECD.

OECD (1960), *Co-Operation in Scientific and Technical Research*, Paris: OECD.

OECD (1962), *Draft 1962 Programme and Budget*, SR (62) 26.

OECD (1963a), *Science and the Policies of Government*, Paris: OECD.

OECD (1963b), *Science, Economic Growth and Government Policy*, Paris: OECD.

OECD (1963c), *The Measurement of Scientific and Technical Activities: Proposed Standard Practice for Surveys of Research and Development*, Paris: OECD.

OECD (1964), *Government and Innovation: Progress Report*, CMS-GI/GD/64/7.

OECD (1965), *The Role of Government in Stimulating Technical Innovation*, CMS-CI/65/63.

OECD (1966), *Government and Technical Innovation*, Paris: OECD.

OECD (1968a), *Gaps in Technology: General Report*, Paris: OECD.

OECD (1968b), *Report of the Conference on Innovation in Education Held in Paris, November 25 and 26, 1968*, CERI/EI/69.19.

OECD (1969), *The Management of Innovation in Education*, Center for Educational Research and Innovation (CERI), Paris: OECD.

OECD (1970a), *Gaps in Technology: Comparisons between Countries in Education, R&D, Technological Innovation, International Economic Exchanges*, Paris: OECD.

OECD (1970b), *Second Draft of the Discussion Paper for Ministers*, DAS/SPR/70.61.

OECD (1971), *Science, Growth and Society*, Paris: OECD.

OECD (1972a), *Ad Hoc Group on Industrial Innovation*, DAS/SPR/72.32.

OECD (1972b), *Analytical Methods in Government Science Policy: An Evaluation*, Paris: OECD.

OECD (1972c), *Innovation in Social Sectors*, SPT(72)8.

OECD (1973), *Case Studies of Educational Innovation, IV – Strategies for Innovation in Education*, Paris: OECD.

OECD (1974a), *Allocation of Resources to R&D: A Systemic Approach*, SPT(74)1.

OECD (1974b), *The Research System*, Volume 3, Paris: OECD.

OECD (1976), *Policies for Innovation in the Service Sector*, SPT(76)12.

OECD (1978), *Policies for the Stimulation of Industrial Innovation*, Volume 1 (Analytical Report), Paris: OECD.

OECD (1980), *Technical Change and Economic Policy*, Paris: OECD.

OECD (1982), *Innovation in Small and Medium Sized Firms*, Paris: OECD.

OECD (1990), *Description of Innovation Surveys and Surveys of Technology Use Carried Out in OECD Member Countries*, DSTI, Paris: OECD.

OECD (1991), *OECD Proposed Guidelines for Collecting and Interpreting Innovation Data (Oslo Manual)*, DSTI/STII/IND/STP (91)3.

OECD (1999), *Managing National Innovation Systems*, Paris: OECD.

OECD (2005a), *An Overview of Public Policies to Support Innovation*, Economic Department Working Paper no. 456, Paris: OECD.

OECD (2005b), *Governance of Innovation Systems*, Volume I: Synthesis Report, Paris: OECD.

OECD (2007), *Innovation and Growth: Rationale for an Innovation Strategy*, Paris: OECD.

OECD (2008a), *Chair's Summary of the OECD Council at Ministerial Level, Paris, 15–16 May 2007 – Innovation: Advancing the OECD Agenda for Growth and Equity*, Paris: OECD.

OECD (2008b), *The OECD Innovation Strategy: Progress Report*, C(2008)62/REV1.

OECD (2009), *OECD Work on Innovation: A Stocktaking of Existing Work*, DSTI/DOC(2009)2.

OECD (2010), *The OECD Innovation Strategy: Getting a Head Start on Tomorrow*, Paris: OECD.

OECD (2014), *Making Innovation Policy Work: Learning from Experimentation*, Paris: OECD.

OECD (2015), *The Innovation Imperative: Contributing to Productivity, Growth and Well-Being*, Paris: OECD.

OECD (2016), *Enhancing the Contributions of Higher Education and Research Institutions to Innovation*, OECD High Level Event on the Knowledge Triangle, Paris, 15–16 September, Paris: OECD.

OECD (2018a), *How Is Research Policy Across the OECD Organised? Insights from a New Database*, Science, Technology and Industry Policy Papers no. 55, Paris: OECD.

OECD (2018b), *Oslo Manual 2018: Guidelines for Collecting, Reporting and Using Data on Innovation*, Paris: OECD.

Ogburn, William Fielding (1922), *Social Change with Respect to Culture and Original Nature*, New York: The Viking Press.

Parsons, Talcott (1956), Suggestions for a Sociological Approach to the Theory of Organizations, *Administrative Science Quarterly*, 1 (1): 63–85; 1 (2): 225–39.

Pavitt, Keith (1963), Research, Innovation and Economic Growth, *Nature*, 200 (4903), 19 October: 206–10.

Pavitt, Keith and Salomon Wald (1971), *The Conditions for Success in Technological Innovation*, Paris: OECD.

Pavitt, Keith and W. Walker (1976), Government Policies towards Industrial Innovation: A Review, *Research Policy*, 5: 11–97.

Perroux, François (1957a), *Théorie générale du progrès économique I, Les composants : 1. la création*, Cahiers de l'Institut de science économique appliquée, no. 59, Paris: ISEA.

Perroux, François (1957b), *Théorie générale du progrès économique I, Les composants : 2. la propagation*, Cahiers de l'Institut de science économique appliquée, no. 60, Paris: ISEA.

Perroux, François (1961), *L'économie du XXe siècle*, Paris: Presses universitaires de France.

Posner, Lawrence D. and Leon J. Rosenberg (1974), *The Feasibility of Monitoring Expenditures for Technological Innovation*, Practical Concepts Inc., Washington, Report to the NSF.

Posner, Michael V. (1961), International Trade and Technical Change, *Oxford Economic Papers*, 13: 323–41.

Presthus, Robert (1962), *The Organizational Society: An Analysis and a Theory*, New York: Knopf.

Price, William J. and Lawrence W. Bass (1969), Scientific Research and the Innovative Process, *Science*, 164, 16 May: 802–06.

Radnor, Michael, Durward Hofler and Robert Rich (eds) (1977), *Information Dissemination and Exchange for Educational Innovations: Conceptual and Implementation Issues of a Regionally-Based Nationwide System*, Center for Interdisciplinary Study of Science and Technology, Evanston (Ill.): Northwestern University.

Reingold, Nathan (1971), American Indifference to Basic Research: A Reappraisal, in Nathan Reingold (ed.), *Science: American Style*, New

Brunswick and London: Rutgers University Press, 1991: 54–75.

Research Management (1970), Top Research Managers Speak Out on Innovation, November: 435–43.

Richland, Malcolm (1965), *Traveling Seminar and Conference for the Implementation of Educational Innovations*, System Development Corporation, report to the Office of Education, Department of Health, Education and Welfare, Washington.

Roberts, Robert E. and Charles A. Romine (1974), *Investment in Innovation*, Report to US National Science Foundation, Midwest Research Institute, Kansas City.

Rosenberg, Nathan (1976), Problems in the Economist's Conceptualization of Technological Innovation, in Nathan Rosenberg, *Perspectives on Technology*, Cambridge: Cambridge University Press: 61–84.

Rossman, Joshua (1931), *The Psychology of the Inventor: A Study of the Patentee*, Washington: The Inventors Publishing Co.

Rostow, Walter W. (1952), *The Progress of Economic Growth*, New York: W.W. Norton.

Rostow, Walter W. (1960 [1999]), *The Stages of Economic Growth: A Non-Communist Manifesto*, Cambridge: Cambridge University Press.

Rostow, Walter W. (1990), *Theories of Economic Growth from David Hume to the Present*, New York/Oxford: Oxford University Press.

Rothwell, Roy and A.B. Robertson (1973), 'The

Role of Communication in Technological Innovation', *Research Policy*, 2 (3): 204–25.

Ruttan, Vernon W. (1954), *Technological Progress in the Meatpacking Industry, 1919–47*, Marketing Research Report no. 59, US Department of Agriculture.

Salomon, Jean-Jacques (1964), International Scientific Policy, *Minerva*, 2 (4): 405–34.

Salomon, Jean-Jacques (1977), Science Policy Studies and the Development of Science Policy, in Ina Spiegel-Rosing and Derek J.S. Price (eds), *Science, Technology and Society: A Cross-Disciplinary Perspective*, London: Sage: 43–70.

Salomon, Jean-Jacques (2000), L'OCDE et les politiques scientifiques, *Revue pour l'histoire du CNRS*, 3: 40–58.

Schatzberg, Eric (2018), *Technology: Critical History of a Concept*, Chicago (Ill.): Chicago University Press.

Schon, Donald A. (1967), *Technology and Change: The New Heraclitus*, New York: Delta Books.

Schon, Donald A. (1971), *Beyond the Stable State: Public and Private Learning in a Changing Society*, London: Temple Smith.

Schumpeter, Joseph A. (1939 [2005]), *Business Cycles: A Theoretical, Historical and Statistical Analysis of the Capitalist Process*, New York: McGraw-Hill.

Sharif, Naubahar (2006), Emergence and Development of the National Innovation System Concept, *Research Policy*, 35 (5): 745–66.

Sherwin, Chalmers W. and Raymond S. Isenson (1969), Project Hindsight: A Defense Department Study of the Utility of Research, *Science* 156, 23 June: 1571–7.

Simone, Daniel V. de (1965), Statement of Daniel V. de Simone, in *Economic Concentration, Part 3: Concentration, Invention and Innovation*, US Senate Hearings before the Subcommittee on Antitrust and Monopoly of the Committee on the Judiciary, Washington: USGPO: 1093–1118.

Simone, Daniel V. de (1967), The Impact of Law on Technological Innovation, in National Science Foundation, *Proceedings of a Conference on Technology Transfer and Innovation*, 15–17 May 1966, NSF 67-5, Washington: National Science Foundation: 37–45.

Simone, Daniel V. de (ed.) (1968), *Education for Innovation*, New York: Pergamon Press.

Simone, Daniel V. de (1974), Trends in Technological Policy Making, in Florence Essers and Jacob Rabinow (eds), *The Public Needs and the Role of the Inventor*, Proceedings of a conference held in Monterey, California, June 11–14, 1973, Office of Invention and Innovation, US Department of Commerce, Washington: USGPO: 17–21.

Skinner, Quentin (2002a), Classical Liberty, Renaissance Translation and the English Civil War, in Quentin Skinner (ed.), *Visions of Politics*, vol. 2, Cambridge: Cambridge University Press: 308–43.

Skinner, Quentin (2002b), Moral Principles and Social Change, in Quentin Skinner (ed.), *Visions of Politics: Regarding Method*, vol. 1, Cambridge: Cambridge University Press: 145–57.

Slichter, Sumner (1928), The Price of Industrial Progress, *The New Republic*, February 8: 316–18.

Smith, Bruce L.R. (1990), *American Science Policy Since World War II*, Washington: Brookings Institution.

Solow, Robert M. (1957), Technical Change and the Aggregate Production Function, *Review of Economics and Statistics*, 39, August: 312–20.

SPRU (1972) *Success and Failure in Industrial Innovation: A Summary of Project SAPPHO*. London: Centre for the Study of Industrial Innovation.

SPRU (1974), *Government Policies towards Industrial Innovation: The final Report of a One-Year Feasibility Study (The Four-Countries Project)*, No editor.

Stamp, Josiah (1929), *Some Economic Factors in Modern Life*, London: P.S. King.

Stamp, Josiah (1933), Must Science Ruin Economic Progress? *The Hilbert Journal*, 32, October–July: 383–99.

Stead, Humphrey (1976), The Costs of Technological Innovation, *Research Policy*, 5: 2–9.

Stern, Bernhard J. (1927), *Social Factors in Medical Progress*, New York: Columbia University Press.

Stern, Bernhard J. (1937), Resistance to the Adoption of Technological Innovations, in US National Resources Committee, *Technological Trends and National Policy*, Subcommittee on Technology, Washington: USGPO: 39–66.

Stevens, Raymond (1941), A Report on Industrial Research as a National Resource: Introduction, in National Research Council, *Research: A National Resource (II): Industrial Research*, Washington: National Resources Planning Board: 5–16.

Tornatzky, Louis G., John D. Eveland, Miles G. Boylan, William A. Hetzner, Elmima J. Johnson, David Roitman and Janet Schneider (1983), *The Process of Technological Innovation: Reviewing the Literature*, Washington: National Science Foundation.

Tornatzky, Louis G. and Mitchell Fleisher (1990), *The Processes of Technological Innovation*, Lexington (Mass.)/Toronto: Lexington Books.

Twiss, Brian C. (1974), *Managing Technological Innovation*, London: Longman.

UNESCO (1971), *International Aspects of Technological Innovation: Proceedings of a Science Policy Symposium, Paris, France 7–9 September 1970*, Paris: UNESCO.

United Nations (1953), *A System of National Accounts and Supporting Tables*, Department of Economic Affairs, Statistical Office, New York.

Usher, Abbott Payson (1929 [1988]), *A History of Mechanical Inventions*, New York: Dover.

US Advisory Committee on Industrial Innovation: *Final Report* (1979), Department of Commerce, Washington.

US Bureau of Labor Statistics (1966), *Technological Trends in Major American Industries*, Department of Labor, Bulletin no. 1474, Washington: USGPO.

US Chamber of Commerce (1966), *Incentives to Private Investment in Technical Innovation*, Washington: Chamber of Commerce.

US Committee on Ways and Means (1980), *Technology and Trade: Some Indicators of the State of US Industrial Innovation*, 96th Congress, Second Session, Washington: USGPO.

US Congress (1955), *Automation and Technological Change*, Subcommittee on Economic Stabilization, Washington: USGPO.

US Congress (1975), *Technology, Economic Growth and International Competitiveness*, Subcommittee on Economic Growth of the Joint Economic Committee, Congress of the United States, Washington: USGPO.

US Council of Economic Advisers (1964), The Promise and Problems of Technological Change, in *Economic Report to the President*, Washington: USGPO: 85–111.

US Department of Commerce (1966), *Technology Transfer and Innovation: A Guide to the Literature*, Office of State Technical Services, USGPO.

US Department of Commerce (1967), *Technological Innovation: Its Environment and Management*, Washington: USGPO.

US Department of Commerce (1972), *Technology Enhancement Programs in Five Foreign Countries*, Office of the Assistant for Science and Technology, Washington: NTIS.

US Department of Defense (1969), *Project Hindsight: Final Report*, Office of the Director of Defense Research Engineering.

US House of Representatives (1980), *Industrial Innovation*, Committee on Science and Technology and Committee on Small Business, Joint hearing before the Committee of Commerce, Science, and Transportation and the Select Committee on Small Business, Part I, Washington: USGPO.

US Joint Economic Committee (1975), *Technology, Economic Growth, and International Competitiveness*, Robert Gilpin, study conducted for the Subcommittee on Economic Growth, Washington: USGPO.

US Joint Economic Committee (1980), *Special Study on Economic Change, Vol. 3 – Research and Innovation: Developing a Dynamic Nation*, 96th Congress, Second Session, Washington: USGPO.

US National Academy of Engineering (1968), *The Process of Technological Innovation*, Washington: National Academy of Sciences.

US National Commission on Technology, Automation and Economic Progress (1966), *Technology and the American Economy*, Volume 1, Washington: USGPO.

US National Resources Planning Board (1941), *Research: A National Resource (II): Industrial Research*, National Resources Planning Board, Washington: USGPO.

US National Science Board (2016), *Science & Engineering Indicators: 2016 Digest*, Washington: NSF.

US National Science Foundation (1956),

Expenditures for R&D in the United States: 1953, *Reviews of Data on R&D* 1, Washington: National Science Foundation.

US National Science Foundation (1958a), Highlights of a Conference on Research and Development and its Impact on the Economy, *Reviews of Data on Research & Development*, 11, NSF-58-25.

US National Science Foundation (1958b), *Proceedings of a Conference on Research and Development and its Impact on the Economy*, NSF-58-36, Washington: NSF.

US National Science Foundation (1959), *Methodological Aspects of Statistics on R&D: Costs and Manpower*, Washington: National Science Foundation.

US National Science Foundation (1961), Research and Development and the Gross National Product, *Reviews of Data on Research & Development*, 26, NSF 61-9.

US National Science Foundation (1966), *Technology Transfer and Innovation*, Proceedings of a Conference organized by the National Planning Association and the National Science Foundation, 15–17 May, NSF 67-5, Washington: NSF.

US National Science Foundation (1972), *Papers and Proceedings of a Colloquium on Research and Development and Economic Growth/Productivity*, NSF 72-303, Washington: NSF.

US President's Commission on National Goals (1960), *Goals for Americans*, New York: Prentice-Hall.

US President's Science Advisory Committee (PSAC) (1964), *Innovation and Experiment in Education*.

US Senate (1965), *Concentration, Invention and Innovation*, Hearings before the Subcommittee on Antitrust and Monopoly of the Committee on the Judiciary, Washington: USGPO.

Utterback, James M. (1974), Innovation in Industry and the Diffusion of Technology, *Science* 183, 15 February: 620–26.

Utterback, James M., Thomas J. Allen, J. Herbert Hollomon and Marvin B. Sirbu (1976), The Process of Innovation in Five Industries in Europe and Japan, *IEEE Transactions on Engineering Management*, 23 (1): 3–9.

Veblen, Thorstein (1915 [2006]), *Imperial Germany and the Industrial Revolution*, New York: Cosimo Inc.

Vernon, Raymond (1966), International Investment and International Trade in the Product Cycle, *Quarterly Journal of Economics*, 80: 190–207.

Warner, Aaron W. (1965), Summation, in Aaron W. Warner, Dean Morse and Alfred S. Eichner (eds), *The Impact of Science on Technology*, New York: Columbia University Press: 197–218.

Watson, Thomas J. (1960), Technological Change, in US President's Commission on National Goals (1960), *Goals for Americans*, New York: Prentice-Hall: 193–204.

Weintraub, David (1932), The Displacement of Workers Through Increases in Efficiency and their Absorption by Industry, 1920–1931, *Journal of the American Statistical Association*, 27 (180): 383–400.

Weintraub, David and Irving Kaplan (1938), *Summary of Findings to Date*, National Research Project on Reemployment Opportunities and Recent Changes in Industrial Techniques, Philadelphia (Pa.): Works Progress Administration.

Wiesner, Jerome B. (1966), Technology and Innovation, in Dean Morse and Aaron W. Warner (eds), *Technological Innovation and Society*, New York: Columbia University Press: 11–26.

Williams, Bruce R. (1956), Science and Industrial Innovation, *The Advancement of Science*, 13 (51): 156–62.

Wisnioski, Matthew (2012), *Engineers for Change: Competing Visions of Technology in 1960s America*, Cambridge (Mass.): MIT Press.

Index